The Selfish Cell

Matteo Conti

The Selfish Cell

An Evolutionary Defeat

 Springer

Matteo Conti
S. Maria delle Croci Hospital
Laboratory of Clinical
Pharmacology and Toxicology
v.le randi, 5
48100 Ravenna
Italy

ISBN: 978-90-481-7941-1   e-ISBN: 978-1-4020-8642-7

Printed on acid-free paper

9 8 7 6 5 4 3 2 1

springer.com

*The martyr sacrifices herself (himself in a
few instances) entirely in vain. Or rather not
in vain; for she (or he) makes the selfish more
selfish, the lazy more lazy, the narrow
narrower.*

*'Forever Yours' (1867). Florence Nightingale
Selected Letters (1989).*

To my family
To everyone who has made even the slightest
sacrifice for me
Thank you for your love

Preface

'I have received in a Manchester newspaper rather a good squib, showing that I have proved might is right and therefore that Napoleon is right, and every cheating tradesman is also right.'[*]

Charles R. Darwin, Letter to Lyell, 4 May (1860).

This book is about cancer, not a new subject by any means. Countless publications are in fact available as a result of as many efforts of scientists, along the years, to exorcise this terrible disease, by recurring to various types of rationalisation, none of which has yet been able to fulfil its goal. The most common type of theoretical framework en-vogue in these days is that of reading carcinogenesis under the lens of Darwinian evolution. This conceptual lens has provided so far, slight improvements in the management of cancer in the clinics but many more in cancer management and prevention.

In this book, we would enter into the details of this view and how it describes the insurgence of evolved selfish inhabitants of nature that are cancer cells. We would therefore discuss around cell selfishness and how it could happen that naïve biological entities, formerly collaborative, at a certain point, turn into parasites prospering at the expenses of their home tissues, paradoxically sealing their own fate by an unsustainable tendency to exploitation.

Such biological selfishness, manifesting in the pathology of cancer, puzzles us far more than any other biological processes under study in our days, because it perturbs our perception of the universe, our innate sense of order and law, at the basis of our lives as social beings. In fact, in cancer, we assist to the insurgency of chaos where order was before, to the victory of anarchy over the law, in a biological metaphor of the rebellion of the Devil against God. In fact, like Satan before being Satan, cancer cells were normal cells, obedient to the rules of nature, in harmony and cooperation with their peers. One day these cells turn into destructive rebels, no more collaborative with their own body of pertinence, aggressing and killing their neighbours without any measure.

---

* from www.quotationspage.com

In other words, in cancer, it appears as if the eternal rules of nature have been broken, sabotaged from inside. In fact, scientists have clearly found that precise biological rules written in the "genetic code" are modified in cancer cells. However, rules, even more vital for the existence of life, are clearly violated during the pathology. With a terminology extrapolated from social sciences, we would indicate them as the tendency to collaboration, to service and to sacrifice for neighbouring components of the same living system.

For some strange willing and since the beginning of history, man has always been inclined to try to disprove or bend similar eternal rules, often ending up in tragedy. Studying the effect of their violation in the world of cells, in this book we would have an occasion to reason over their value, as a source of evolution, as opposed to competitive tendencies, that are corroding our social tissue in western societies as well as natural environment.

Furthermore, for some strange mental habit we start fully noticing the value of things when we are about to lose them. Therefore, observing the process of carcinogenesis spoiling living tissues and systems, we would have an occasion to be reawakened to the harmony and equilibrium that pervades them.

Nature is all dominated by this amazing intrinsic harmony; the bodies of living creatures being just a mirror of that. Even if poetry is undoubtedly the eligible discipline to describe such a miracle of orchestration, with the arid words of science, we can just tell the tale of an amazing cooperation of cells, all daughters of a single embryonic cell; working together to build a higher-order form of existence in what appears an incredibly complex integration of molecules, genes, proteins, former micro-organisms, later transformed into organelles.

In front of a similar construct, of integration, which is the root of life itself and self-evident in evolved life forms, one could become persuaded that, if evolution has ever proceeded in competitive terms, collaboration should have become competitively advantageous at some point in history, if not from the beginning of times. The hope is that the study of carcinogenesis, interpreted with that key of lecture as provided in this book, could become a warning for scientists, and for humans in general, to finally stop opposing and instead trying to harmonise to the rules that constitute and govern our universe as a whole.

As a last introductory note to the reader, we must state clearly that the analysis of the phenomenon of cancer in this book is subjected to conceptual filters currently available to scientists in our days. We are aware that this field of biological research is still in active ferment and that on this background, open questions are probably more valuable than comprehensive explanations. This is why we will try to forward ideas more than explanations, with the hope that they would be of some value for researchers. We wish that they could work with an open mind, free of the bounds of money or the constraint of currently accepted models. Only in this way, cancer research would become an opportunity to discover more of the wonders of create and to learn important lessons for mankind.

Further Readings

Singer, P., A Darwinian Left. 1999, London: Weidenfeld & Nicolson.

Robert G. McKinnell, R.E.P., Alan O. Perantoni, G. Barry Pierce, The Biological Basis of Cancer. 1998, New York: Cambridge University Press.

Nature milestones in cancer. Nat Rev Cancer, 2006: S1–Sxxx.

Dawkins, R., the selfish gene. 1989: Oxford University Press. 366.

Velicer, G.J., Evolution of cooperation: does selfishness restraint lie within? Curr Biol, 2005. 15(5): R173–5.

Weinberg, R.A., The Biology of Cancer 1ed. 2006, London: Garland Science.

Weinberg, R.A., One renegade cell: how cancer begins. 1998, New York: Basic books.

Contents

Chapter 1
Selfish Cells

> *'The greatest obstacle to discovery is not ignorance - it is the illusion of knowledge.'*
>
> *Daniel J. Boorstin (1984)*.*

Schema huius præmiſſæ diuiſionis Sphærarum.

The Tolemaic geocentric model of the Universe. Because of his views about cosmology, Galileo Galilei was persecuted, imprisoned and had to swear that he was wrong. Despite persecution the truth emerged with time

---

* www.quotationspage.com

Introduction

What is cancer, at least in scientific terms? If challenged with a direct question, I think many of us in the field could be tempted to sneak away with one or another shortcut answer, even if we know full well that a thorough definition is not actually available so far to the best of our knowledge. However, to get the issue started I would begin with a standard conceptualisation, that could be retrieved on internet sites of major health organisations or in biomedical books and journals.

> 'Cancer is a group of diseases in which cells are aggressive (grow and divide without respect to normal limits), invasive (invade and destroy adjacent tissues), and/or metastatic (spread to other locations in the body). These three malignant properties of cancers differentiate them from benign tumors, which are self-limited in their growth and do not invade or metastasize (although some benign tumor types are capable of becoming malignant)...'.*

According to similar definitions, cancer is a disease caused by the own cells of an organism. Cancer cells are therefore commonly envisioned as parasitic life forms. In fact, they exhibit a parasitic life-style made of growing and reproducing at the expense of their hosts.

In this chapter we will try to enter a bit more into the concepts behind similar definitions, to set the stage for a discussion that would take most of the next chapters of the book to be completed. I would only anticipate that, despite cancer research has today lead the assessment of the disease to its molecular terms, such simple empirical definitions most probably holds a key to enter a deeper level of understanding of the phenomenon that could even enable us in the end to fight it off in our lives.

Observations

In the ancient Greece, with the tools to him available, Hippocrates observed and consequently described several types of cancer. Benign tumours he termed oncos, the Greek word for swelling, whereas malignant tumours he termed carcinos, the Greek for crab. The name probably was assigned due to the superficial appearance of some solid malignant tumour, with a hard center surrounded by malignant tissue projections, vaguely resembling the shape of a crab. In morphological descriptions, it was often used the suffix -oma, again meaning swelling in Greek, thus originating the current term carcinoma.

Celsus translated carcinos into the Latin cancer, also meaning crab, coining the word we use today. Well later, Galen dusted off the term oncos, to describe generically all tumours, founding the etymological root for the modern word oncology.

Without much insights of the cellular and molecular composition of living matter, picturesque descriptions characterised oncology until 16th and 17th century, when the eminent German professor Fabry believed that breast cancer was caused by milk clots in a mammary ducts, or the Dutch professor Sylvius, a follower of Descartes,

---

* Wikipedia

sustained that cancer was the outcome of retention of acidic lymph fluids. Accordingly, patient's treatment consisted of diet, blood-letting and laxatives. These are actually supportive measures, also in modern approaches, but not usually sufficient to cure the disease.

Similar humor-based theories remained popular until the 18th century, when finally the widespread use of the microscope enabled scientists to observe the fundamental building block of biological tissues. This was a real turning point, a quantum leap to almost where we are now.

Microscopy examinations, first of normal tissue sections and later of sections made from tumor masses revealed that tumors, like normal tissues, are composed of cells. And since normal cells constitute normal tissues, Rudolph Virchow, one of the fathers of histology, postulated that abnormal tissues (tumors) are composed of cells that have, in some way, become abnormal in nature. Actually, to define this process of cell alteration, Virchow used the term degeneration. A term most often used even in these days.

Only at the end of the 19th century, the English surgeon De Morgan provided the first cell based description of the phenomenon of metastasis, describing cancer cells spreading from a primary tumor to other tissues where they form multiple other tumors.

Today histopathologists, examining the microscopic appearance of tissues, classify tumors elaborately. It is not the goal of this book to thoroughly summarize the complex system of classification of cancer histopathology. However, just to provide a general framework, for those who are out of this field, we would mention a few general concepts.

Current classification groups tumors into two broad categories, depending on the degree of their aggressiveness: those that grow locally without invading adjacent tissues (benign) and others invading nearby tissues and spawning metastases (malignant). Most of the primary tumors are benign and can become dangerous only occasionally, when their expansion press on vital organs, or when they secrete substances (principally hormones) that create physiological imbalances in the body. Today, cancer-related mortality derives mainly from malignant tumors and metastases are responsible for about 90% of deaths from cancer.

Tumors are classified into four major groups, according to their presumed histological origin. These are epithelial (carcinomas), mesenchymal (sarcomas), haematopoietic (leukaemias and lymphomas), and neuroectodermal tumors. Occasionally, the term anaplastic is used to refer to masses in which their cells are lacking tissue-specific traits. At times, histologists also annotate cells of unclear origin in certain tumor masses as, for instance, the often reported neuroendocrine secretory cells in small-cell lung carcinomas (SCLCs).

The great majority human cancers are of epithelial origin and are thus carcinomas. Among the human carcinomas are tumors arising from almost every epithelia cell layers in the body: the mouth, esophagus, stomach, small and large intestines, skin, mammary gland, pancreas, lung, liver, ovary, urinary bladder, etc.

Fortunately, hystologists do not always report of carcinomas. They often observe in surgical biopsies only pre-malignant tissue morphologies, meaning altered tissues that have not a fully tumoral aspect. These are classified in order of progressive cellular disarray in the tissue as hyperplastic, metaplastic, dysplastic tissues. Hyperplasia is the first grade of disorder, considered normal except for an excessive number of cells. Metaplastic tissues show already displacement of cells due to the action of cell types not usually encountered at those sites. Dysplastic tissues show instead also cells that are cytologically abnormal. Adenomatous growths, such as adenomas, polyps, papillomas, and warts are dysplastic tissues, considered to be benign until they respect the boundary created by the basement membrane of an epithelium. Tumors that breach the basement membrane and invade underlying tissue are classified as malignant. In fact, the breaching of the basement membrane is a crucial step toward metastasis. This is the worst situation, with seeding of cells to form tumor colonies to other sites in the body. This disparate collection of benign and malignant growths is collectively referred to as neoplasias.

Because of the existence of increasing degrees of tissue abnormality, scientists consider likely that these are steps along the road that leads from normal tissues to malignant ones. However, it must be emphasized that histological examination is based on biopsies, on the removal of similar intermediate states of degeneration and not on their real-time progressive observation. Therefore doubts remain about the actual steps of malignant progression of tissues toward cancer. Live microscopic examinations have not often been technologically feasible, so far, except in a few specific cases of epidermal animal tumors that, for what it matters, confirm the general overview. A more detailed discussion of cancer progression is provided in Chapter 3.

For the sake of exactness, it must be said that histological classification is only rarely as simple as it would be indicated in the depiction above. In fact, the situation under the eye of the pathologist is usually much less defined. Only discarding many details, classic straightforward definitions can be achieved.

As it usually happens, details make the difference, and clinical routinely descriptions, useful as they are in dictating therapeutic strategies, today too often miss crucial aspects of tumor appearance and lose therefore invaluable opportunities to improve therapeutics approaches that could be at reach employing, for instance, molecularly targeted fluorescent probes available to researchers.

If we take, for instance, a crucial changes in cell phenotype, such as the epithelial-mesenchymal transition (EMT), despite its importance in pathogenetic terms, it is often reported only in research laboratories and very seldom in clinical laboratories, where the great mass of specimens are processed instead. EMT apparently accompanies and enables the invasion of carcinoma cells into adjacent normal tissues and could be often observed at the borders of carcinomas. However, the EMT is a phisiological and and often useful phenotypic change that epithelial cells should perform in the reconstruction of cell layers after wounding or in certain morphogenetic steps occurring during embryogenesis.

In metastasis, cancer cells are retrieved at distant sites in the body, in tissues very different from those of origin, at least accordingly to cytological considerations. Apart from simplifying definition, our understanding of metastasis is still rather incomplete, explaining why metastasis represent a major danger for health.

Usually, the metastasis process is seen as a gradual series of progressive steps, and is therefore divided into the following, self-explicatively termed, phases: local invasion, intravasation, transport, extravasation, formation of micrometastases, and colonization.

Initially, in carcinomas, cells are confined to the epithelial compartment by the basement membrane. This is considered a dangerous situation, but not a lethal one, as long as the cells remain histologically localised. However, as already mentioned, carcinomas cells can acquire the ability to breach the basement membrane, and begin to invade the nearby stroma, therefore starting off a local invasion.

If cell viability would depend on anchorage to extracellular matrix of origin, as it is common for epithelial cells, the migrating cells would rapidly die from anoikis, a form of apoptosis triggered by detachment from solid substrates. Instead, carcinoma cells manage to survive the detachment and entering in contact with the stroma that seems to benefit them in multiple ways, by supplying mitogenic and trophic factors. Cancer cells show various degree of dependence on "normal" neighbor cells, for various types of support in malignancy. The crucial influence of the stroma will be better defined in Chapter 4.

Once in the stromal compartment, cancer cells can reach the blood and the lymphatic system vessels. This is when intravasation takes place. Pro-invasive properties are needed to move through the capillary walls and into the blood. Once in the circulation, metastasizing cancer cells can attract an entourage of blood platelets that shields them by the mechanical stress of blood flow and by attacks of the immune system. In fact, in cancer patients, an abnormal coagulation of the blood is often observed that is a consequence of the secretion of procoagulant factors from cancer cells. Disseminated intravascular coagulation (DIC) is one of the major cause of death related to the presence of a malignant cancer.

The blood circulation is a preferential way by which cancer cells can travel. Anyway, almost all tissues in the body carry networks of lymphatic vessels that cancer cells can use to reach distant sites. Histological analyses of spontaneously arising tumors indicate that lymphatic vessels are usually collapsed or broken due to their lack of mechanical resistance to cell crowding and consequent increase in mechanical pressure inside tumor masses. Nevertheless, some cancer cell does succeed in entering viable lymphatic vessels, where they can be easily detected by histologists, because their appearance differs so strongly from the surrounding lymphoid cells.

Histological examination of draining lymph nodes is routinely used to determine whether a primary breast cancer has begun to dispatch metastatic cells to distant sites in the body. Lymph nodes examination represent a useful "surrogate marker" of metastasis by providing useful diagnostic and prognostic data, without being directly involved in the processes that lead to widespread cancer cell dissemination and metastatic disease.

Metastasizing cells in the bloodstream encounter natural filters constituted by the capillary bed of the lungs. Most cancer cells, having tens of microns diameter size plus an additionally cloak of platelets, are too big and irregularly shaped for passaging through most capillaries, whose internal diameters is only in the range of a few microns. Trapped within the lungs, cells may attempt to metastasize onsite.

However, metastases are often found elsewhere in the body, indicating that cancer cells frequently succeed in escaping from the lungs and travel further to other sites in the body.

During extravasation they escape from the lumina of vessels and penetrate into the surrounding tissue. To surface from the blood circulation, cancer cells can exploit specific receptors, for instance integrins, that mediate the adherence to the luminal walls of arterioles and capillaries. Vascular beds of specific organs can express specific patterns of adhesion molecules on their surfaces that can attract cancer cells.

Arrived within the parenchyma, metastasizing cancer cells can colonize the new tissue. However, before the attack, they may survive for extended periods of time in small clumps of a few cells. These are micrometastases, that can be disseminated potentially throughout the tissues of cancer patients.

Support for the existence of dormant micrometastases, which persist in a non-growing state for extended periods of time, comes from experiments with dye-labeled cancer cells, artificially introduced via the portal circulation into mice liver. Weeks later, cancer cells can be recovered from livers in a viable but quiescent state. Micrometastases represent a serious threat, because they may erupt years after a cancer has been judged clinically cured.

At the end of the 19th century, the British pathologist Stephen Paget reported that specific organs were preferential targets for metastasis of specific primary tumors. In his now famous 'seed and soil' analogy, he compared cell dissemination with the dispersal of the seed of plants:

> 'A plant goes to seed, its seeds are carried in all directions; but they can only live and grow if they fall on congenial soil'.

In general, the frequency of metastases to an organ is governed both by the frequency with which metastasizing cells are physically trapped and the ease with which they can adapt to the microenvironment of an organ; but, with much more data to us available today, we are aware that the "seed and soil" hypothesis cannot explain the metastatic patterns of all types of human cancers.

There are also important factors depending on transient pathological and even physiological states of the body, specially due to localized wounding and chronic inflammatory states of tissues, that provide a spectrum of mitogenic and trophic signals, active secretion of hormonal substances, chemo-attractants (chemokines) that might actively call for wandering cancer cells to enter specific tissues.

Similar mechanisms mediated by heterotypic cell signalling, involving multiple cell types, are recently being investigated as critical modulators of cancer phenotype and its progression in general, in addition to their role in the metastatic cascade mentioned above. In fact, advanced microscopic imaging techniques have enabled simultaneous observation of various cell constituents in biological tissues. With the aid of these new tools, we are progressively becoming aware that selfishness is not a peculiarity of cancer cells alone and that other cell types, in particular stromal cells and immune cells, assist the process of cancerogenesis.

In these days, in addition to the discovery of those contribution in cancer progression, a new series of evidences are being collected on the existence of previously unknown cancer stem cells. These discoveries in particular are inducing a serious review of current models of tumorigenesis. We will come to that issue more in detail in the next chapters. Here, briefly, we anticipate that tumors have come to be considered organized tissues, much like normal tissues, in which self renewing stem cell population is responsible for spawning the bulk population, while the rest are 'transit-amplifying cells'. Cancer stem cells would be true tumorigenic cells, in that they can seed new tumors when implanted into a host, whereas non-stem cells (the transit-amplifying cells) would be unable to do so, because they can only perform a limited number of replications. Therefore, the ability to create macroscopic metastases would be related to the proliferative potential of cells that have escaped the primary tumor. Cancer stem cells having an unlimited ability to replicate could be ideally suited to spin off metastatic colonies that would expand to life-threatening size.

Undifferentiated stem cells, having access to the complete range of biological functions classically not accessible to differentiated cells, should easily have access to those phenotypic features necessary to perform the metastatic phases described above, switching on and off specific functions at the right moment in response to environmental stimuli. Cancer cells could activate morphogenetic programs that are encoded in their genome from the beginning, and whose expression depends on the environment or on peculiar stimulations received by other cells. These are multiple changes in cell phenotype, dynamically modulated in ways that they can not be genetically acquired, as it was thought in the past. In the light of novel discoveries, a reclassification of almost a century of experiments is under course, which is the subject for a later chapter of this book.

Further Readings

nature milestones cancer. Nat Rev Cancer, 2006: S1–Sxxx.

Birchmeier, C., et al., Met, metastasis, motility and more. Nat Rev Mol Cell Biol, 2003. 4(12): 915–25.

Brabletz, T., et al., Opinion: migrating cancer stem cells – an integrated concept of malignant tumour progression. Nat Rev Cancer, 2005. 5(9): 744–9.

Chambers, A.F., A.C. Groom, and I.C. MacDonald, Dissemination and growth of cancer cells in metastatic sites. Nat Rev Cancer, 2002. 2(8): 563–72.

Condeelis, J. and J.E. Segall, Intravital imaging of cell movement in tumours. Nat Rev Cancer, 2003. 3(12): 921–30.

Nguyen, D.X. and J. Massague, Genetic determinants of cancer metastasis. Nat Rev Genet, 2007. 8(5): 341–52.

Reya, T., et al., Stem cells, cancer, and cancer stem cells. Nature, 2001. 414(6859): 105–11.

Thiery, J.P., Epithelial-mesenchymal transitions in tumour progression. Nat Rev Cancer, 2002. 2(6): 442–54.

Chapter 2
The Seed of Selfishness

'This is the great truth life has to teach us ... that gratification of our individual desires and expression of our personal preferences without consideration for their effect upon others brings in the end nothing but ruin and devastation.'*
Hortense Odlum. A Woman's Place (1939)

Guernica (Pablo Picasso, 1937). The painting depicts the Nazi German bombing of Guernica, Spain, on April 26, 1937 during the Spanish Civil War. Guernica depicts suffering people, animals, and buildings wrenched by violence and chaos

Introduction

Thanks to the empirical accounts of hystologists, in the first chapter we have depicted cancer cells growing abnormally, in high numbers, invading tissues. These are selfish cells, meaning that they do not respect their boundaries, imposed by the presence of other cells; work incessantly to span their progeny at the expense of the organism in which they reside. These cells were formerly normal cells, respectful of their role in the community of their peers, forming a biological tissue.

---

* From www.brainyquote.com

Observations of biologists, biochemists, epidemiologists and researchers in general, during the past century, looking for the causes of that deranged behaviour of cells, have conveyed in the view that degeneration of cells to a cancerous state is to be attributed to abnormalities in their genetic material, caused by the exposure to chemicals, electromagnetic high energy radiations or infectious agents.

In these searches, scientists were more or less consciously influenced by a paradigm that we should state clearly before going on: that the phenotype of those cells, and of life forms in general, is simply determined by their genotype. Holding to that paradigm, we can coherently proceed with the content of this chapter; putting for the moment in background those relatively recent notions in biology that are inducing scientists to reconsider that pilaster of biology as simplistic, in these days. In fact, novel aspects of modern molecular biology, like epigenetic alteration of genes, the role of non-coding regions of eukaryotic DNA (mostly not yet deciphered), of micro-RNAs, as well as elements related to tissue hierarchical organisation and stem cells, would hopefully soon lead to a more proper description of life science phenomena, and consequently of cancer as well.

Indications from Epidemiology

Epidemiology has provided with substantial indications about where to look for the seed of cancer. This relatively recent discipline has essentially started only in the 20th century, when it became possible to observe and statistically evaluate significant groups of oncology patients, after massive hospitalisation induced by the World Wars.

Early population based observations, however, had already been published in the literature well before that period by pioneers of the field, like the British surgeon Percival Pott, who had observed in 1775 that the rare cancer of the scrotum was instead a common disease among chimney sweeps in England. Data on the correlation between exposure to electromagnetic radiations and cancer incidence, appeared in the first decade of the twentieth century, with studies on the use of the recently invented X-rays, had shown that people working with those radiations were subject to increased rates of cancer, often developing tumors at the site of irradiation.

A founding report is that of Janet Claypon who published in 1926 a comparative study about 500 breast cancer cases and 500 control patients of the same background and lifestyle for the British Ministry of Health. However, fathers of the discipline are considered Richard Doll and Austin Bradford Hill who published "Lung Cancer and Other Causes of Death in Relation to Smoking, and an additional "Report on the Mortality of British Doctors" in 1956. Doll was the founder of the Oxford unit for Cancer epidemiology in 1968. His team, with the use of early 'powerful' computers, is considered the first to compile really significant amounts of data in cancer epidemiology.

Modern epidemiology has later greatly developed due to its links with public health policies. Thousands of studies have been performed and strong conclusions on cancer aetiology finally drawn from them. Epidemiology has clearly assessed, for instance, that the incidence of some type of cancer (mostly pediatric ones) is comparable worldwide but specific types of neoplasias varies dramatically from a population to another. From similar observations, it has been discriminated between the two main factors determining cancer rates in various populations: heredity and environment. Many studies have, in fact, contributed to define that the great majority of cancer risk is environmental, or related to lifetime exposure. In the Japanese population, for instance, rates of stomach cancer are extremely high, about ten times higher than those of Americans. However, Japanese settled in the United States, within a generation, exhibit rates comparable to that of the general population.

That the environment is the main determinant of cancer incidence has been strongly confirmed by population based studies on lung cancer rates in non developed countries. It has been observed an increase in the disease rate of this disease, in parallel with an increasing smoking habit of those populations. In addition, within genetically homogeneous populations, there are vast differences in cancer rates depending on lifestyle habits. American Adventists, for instance, whose religion discourages smoking, heavy drinking, and the eating of meat, die from cancer at a rate half of that of the general population. The general indication extrapolated from similar studies is that the great majority of commonly occurring cancers are caused by factors that are external to the body and after entering into the body manage to corrupt it.

Similar conclusions can be drawn from observations on cancer incidence in animals, even if reliable population-based statistics are much more fragmentary than in humans. In general, it has been observed that a propensity to develop tumours is a common feature of multicellular animals but, even if cancer rates might be expected to escalate with increasing complexity and longevity, homo sapiens seems to experience a substantially greater risk than other animals, including the great apes.

Accordingly, on captive or domesticated animals, veterinary pathology indicates a modest although significant rate of both benign and malignant cancers, compared with corresponding wild types. There are some striking and informative examples of animals with high rates of cancer, all related to human intervention due to inbreeding and ageing in captivity, inbreeding and unnatural reproductive or growth histories, exposure to man-made carcinogens, as in the famous case of the whales of St. Lawrence estuary, ovarian cancer in battery hens exposed to constant light, and mammary and uterine cancers in captive felines given contraceptives. All these data hints to the fact that cancer risk is greatly exacerbated by certain elements connected to peculiar aspects of human activity.

Indications from Chemical and Physical Carcinogenesis Studies

Some of those cancer inducing exogenous elements suggested by epidemiology have been individuated by means of studies of chemically induced carcinogenesis, in laboratory animals. Experiments with coal tar, for instance, performed in Japan by Katsusaburo Yamagiwa at the beginning of the twentieth century, showed that carcinomas in rabbits could be induced, after many months, by repeated painting of localized areas of their ears' skin. Products of combustion, such as 3-methylcholanthrene, benzopyrene, dibenzanthracene and many others, that can be found in the condensates of cigarette smoke as well, are the actual carcinogens in coal tar. To be exact, their metabolites, more than the original molecules, are the real carcinogens.

Independent experiments performed on *drosophila* flies by geneticists indicate that the same types of compounds are mutagenic, leading to the idea that the property of causing DNA mutations could be linked to carcinogenesis. The hypothesis was vastly investigated during the 70s, by means of in vitro methods, like that invented by Bruce Ames at the University of California Berkeley. Ames developed a system to rapidly perform a screening of the mutagenic potential of chemicals, based on their ability to mutate the bacterial genome. The experimental protocol consisted of applying the compounds to *Salmonella* strains growing on Petri dishes and observing the development of eventual mutants, evidenced by the ability to live into antibiotics containing media, thank to an acquired genetic resistance. Ames's test was used to demonstrate that many known carcinogens are also mutagens and that their carcinogenic power directly correlates with their ability to induce mutations in bacterial cells.

As researches on mutagens proceeded, it became apparent that virtually all compounds that are mutagenic for the DNA are likely to be carcinogenens, as well. The same conclusion was strongly supported by the results of studies on X-rays that are recognised both as mutagens and carcinogens. This convergence of results apparently offers a unifying theory of carcinogenesis, in which substances or radiations are carcinogenic through the ability to injure the DNA of cells.

Indications from Infectivology: The Discovery of Oncogenes

Not only chemicals and radiations, but even infectious agents are able to induce cancer in multicellular organisms. This fact is long known since old reports, back in 1908 in Copenhagen, about filterable agents from chicken leukaemia that could transmit the disease to other birds. Peyton Rous in 1910 succeeded in transmitting sarcomas among chickens by preparing small fragments of tumors and implanting them into syngeneic animals. Rous was also able to induce cancer by injections of the same tissue ultrafiltrates, therefore containing viral sized particles later isolated and named Rous sarcoma viruses (RSVs). Soon thereafter, many other reports confirmed the existence of oncoviruses, like a transmissible virus in rabbit myxomas, two more chicken oncoviruses found by Rous's group or different sarcoma viruses discovered in Japan.

However, it has also happened historically that infectious agents have been mistakenly attributed with carcinogenic properties. Many of the reader would probably know the episode occurred in 1926 to the Danish researcher Johannes Fibiger, who was also awarded with the Nobel Prize for his researches on carcinogenic worms, able to induce stomach cancer in rats. A few years later, researchers scrutinizing his experiments, discovered that Fibinger's rat neoplasms were not true cancers, but hyperplasias caused most probably by vitamin deficiencies. After that accident, for more than 30 years, very few authors hazarded to report novel infectious agents able to induce cancer.

Most of what we know about infectious agents as etiological agents in cancer is due to the wave of discoveries that rushed during the 70s, taking the lead from encouraging results obtained by Howard Temin and Harry Rubin's in Berkeley. They invented protocols to transform chicken fibroblasts by direct infection with RSV virions in vitro. The transformed cell phenotype was found to be retained by infected cells, even transmissible to progeny in culture. It was assessed that the retention of the transformed state was due to a single product of the viral gene named *src*, required to both initiate and maintain the transformed phenotype. *Src* is considered the first oncogene discovered.

Following the track of this virally carried oncogene, the greatest discovery in the field of oncology took place in 1975, when Michael Bishop and Harold Varmus published that the genetic sequence typical of the *src* gene was present in the DNA of uninfected bird cells. Later it was retrieved also in other vertebrate species, as well as in mammals. The presence of highly conserved *src* sequences in the genome of so many organisms implied that *src* is a fundamental player in all cells. The concept was revolutionary: it implied that the genomes of normal cells carry genes that have the potential, under certain circumstances, to turn into malignant ones. Bishop famously said, that 'the seeds of cancer are within us'.

Soon it was realised that the corresponding viral derived *src* transforming gene (afterward termed *v-src*) was closely related to the cellular *src* gene (*c-src*) but not exactly identical. It was demonstrated that during the course of infection, a virus can steal genetic sequences from its host and incorporate them into its own genome. Viruses have the ability to exploit similar pre-existing cellular genes for their own purposes. In fact, the stolen sequences, mutated by genetic hypervariablity mechanisms can be converted into genes capable of driving cell proliferation. Infecting new hosts the virus ends up horizontally spreading the bad seed of cancer in the world of cells.

The reality of oncogenes offered a unifying cause for physical, chemical and infectious carcinogenesis. All carcinogenic agents could similarly induce tumors through their ability to mutate critical cell genes, termed proto-oncogenes into oncogenes, the bad seed of cancer.

Historically, DNA and RNA tumor viruses suggested to cancer biologists a simple and powerful theory on how human tumors could arise.[1] However, most of the attempts to isolate cancer viruses from most types of human tumor biopsies failed.

---

[1] All these experiments on viruses opened the door to the understanding of cell transformation at a molecular level. However, their success contributed to support the illusion that the biology of tumors could be completely understood from studies of cells in vitro. Today we know that this is probably not true, since it is becoming increasingly clear that cancer is a disease of malfunctioning tissues rather than one of malfunctioning of cells.

In addition, when tissue ultrafiltrates obtained from tumors infected by viruses were used in transfection experiments, researchers were unable to induce a transformed state in human somatic cells in vitro.[2] Finally, epidemiologic analyses revealed that human cancers do not spread from one individual to another as an infectious disease. Therefore, the role of infectious agents in carcinogenesis was discredited in the past. Recently their role has been reconsidered thanks to new evidences that viral infections are causing a variety of human tumors, some of them quite common.

Today, we estimate that about one-fifth of deaths from cancer worldwide are associated in one way or another with viruses. Many carcinomas of the liver (hepatomas) have been attributed to chronic hepatitis B (HBV) and hepatitis C (HCV) infections. The search for DNA viruses in humans has lead to the recently confirmed discovery that almost all human cervical carcinomas carry fragments of human papillomaviruses (HPVs) genome integrated in their chromosomes. The evidence the role of HPV in this case is so strong that vaccination against HPV is nowadays recommended for the prevention of these cancers worldwide, to the extent that various national health campaigns have been started.

In many cases, infecting viruses carry oncogenes with them, much like animal viruses mentioned previously. This has been observed for particular strains of Epstein-Barr virus (EBV), of Kaposi's sarcoma herpes virus (KSHV also named HHV-8), of certain strains of HPV, of human T-cell leukemia virus (HTLV-1), of polyomaviruses SV40,[3] JCV and BKV.

Other viruses, such as the famous HBV, HCV and certain HPVs mentioned above, do not apparently carry significant oncogenes with them.[4]

Similarly, other carcinogenic infectious agents, in particular bacteria, involved in cancer causation, like *Helicobacter pylori* in gastric carcinoma and lymphoma and *Salmonella typhi* in hepatobiliary carcinomas; and also in other cases of parasite

---

[2] Today scientists are beginning to reconsider those in vitro experiments on human cell transformation, in the light of recent concepts about stem cells and differentiation. In fact, in those experiments with viruses researchers used human somatic cells, in advanced stages of differentiation that we know to be very refractant to transformation. What would be the result of similar experiments targeting less differentiated cells with oncoviruses?

[3] The SV40 monkey virus became famous because of millions of doses of polio oral vaccines, administered to humans between 1955 and 1963, contained the virus incidentally and more than 10 million people were exposed to it. Fortunately, epidemiologic analyses conducted later indicated little increases of cancer cases among those who had been vaccinated.

[4] We should probably mention at this regard that human adenoviruses, known to be responsible for upper respiratory infections, are able to induce tumors when administered to hamsters. The virus can multiply freely in its natural host cells, which are permissive for viral replication, but when introduced into non-permissive cells, such as those of small animals, they fail to replicate and leave behind transformed clones. Similarly, herpesviruses of the Saimiri monkeys are able to provoke rapid and fatal lymphomas into other monkey species, confirming the view that non-permissive cells can become transformed when infected by viruses. This indirect ability to induce cancer is probably to be attributed to the ability of the infectious agent to induce chronic tissue damage and inflammation, an issue that will be covered later.

association with human cancers, such as those of nanobacteria and of mycoplasmas, the presence of oncogenes in their genome has never been recognized, thus far.[5]

However, an intriguing set of data indicates the existence of a back-flow of genetic information from human cells toward infectious agents. In particular, strains of common bacteria, such as *Staphylococcus*, living in close contact with transformed cells in vitro, have been observed to be able to steal eukaryotic genes and use them for their purposes; much like cancer retroviruses seen above. The possibility that these common bacteria could acquire oncogenes during their infection of human cells and could turn them backwards to cells in further infection episodes has not yet been verified nor ruled out.

This evidence clashes with the paradigm of modern oncology that predicts that carcinogenic information, both virally received or acquired by mutation, is naturally unable to get out of cancer cells, in other words that oncogene transmission can be exclusively vertical among cells. This common view is widespread despite not being supported by experience. Actually, it is even contradicted by the same existence of oncoviruses.

In addition, horizontal transfer of oncogenes in vitro is known since the late 70s, with the advent of gene transfer procedures. In particular, by means of the electroporation technique, it was possible to transfer DNA fragments from cancer cells into normal recipient cells in vitro that became consequently transformed. In these experiments, cells derived from mouse embryo fibroblasts were mainly used as acceptors, because they were especially efficient in receiving and integrating foreign DNA.[6] These transfection experiments suggested that oncogenes could be transferred to the genomes of cells when cell protective barriers were lowered, in that case by applying a very drastic and a specific technique named electroporation.[7]

In the literature, horizontal transfer of oncogenes in vivo has been also observed to take place by the direct uptake of DNA via apoptotic bodies in phagocytic competent cells, such as immune cells, fibroblasts and endothelial cells. Nature has provided cells with protective mechanisms that are able to sense degraded DNA, deriving from apoptotic bodies, hence preventing these genes from being replicated.

---

[5] Transmission of oncogenes is known to take place only in one case: that of tumorigenesis caused by the bacterium *Agrobacterium tumefaciens* in dicotyledonous plants. However, similar mechanisms of carcinogenic information transfer have never been recognized in bacterial associations with human tumors.

[6] DNAs extracted from cell lines derived from human bladder, lung, and colon carcinomas, as well as from leukemia, were all capable of transforming recipient mouse fibroblasts. This also suggests that the oncogenes in these cell lines are capable of acting across species and tissue boundaries to induce cell transformation.

[7] Very recently some author has advanced an hypothesis that fusogenic viruses could be carcinogenic through their ability to promote cell membrane fusion events, similarly to what happens with electroporation. Their carcinogenic effect could even be independent on the presence of eventual oncogenes, contained in the nuclei of formerly single cells, because karyotypic disorder following fusion would be *per se* carcinogenic.

However, in case of malfunctioning of these defensive mechanisms, the accumulation of dangerous genetic material could contribute to the spreading of carcinogenic information in the world of cells.

This view has been taken to extremes and a possible role of tumor derived nucleic acid fragments in the development of remote metastasis has been advanced in the literature. Specifically, the hypothesis known as the 'genometastasis', proposes that cells might be naturally transfected with dominant oncogenes as a result of dissemination of such genes in blood of cancer patients. Accordingly, even low probability transformation events, such as those hypothesised for transfection of cells with naked circulating nucleic material, could have a role in cancer spreading in patients. In most cases, however, protective mechanisms are much likely to impede this phenomenon to take place.

In addition to horizontal spreading of oncogenes among eukaryotic cells, there is a dangerous way which open the door to oncogene 'infections' in vertebrates: germ line transmission. This situation has been clarified in a series of well known experiments in which mice or chickens were exposed to retroviruses that resulted in the infection of a wide variety of cell types in their body including, on occasion, the cells in the gonads. Infections of the germ line resulted in the integration of viral genes into the chromosomes of precursors to either sperm or egg. When these gametes participated in fertilization, the viral genes could be transmitted to a fertilized egg and thus to all of the cells of a resulting embryo and later of the adult. These endogenous viral genes are readily transcribed in animal's tissues creating viremia and inducing cancer early in life.

Careful examinations of the genomes of a variety of mammalian and avian species have demonstrated the presence of numerous endogenous retroviral genomes. Most have suffered so many mutations that they are no longer able to specify infectious viral particles for they have probably resided for millions of years in vertebrate germ line. However, a small subset of endogenous viral genomes, notably those that have been recently inserted into a species' germ line, could remain genetically intact. Given the proper stimulus, these previously latent genes may suddenly be transcribed in one or another cell, and eventually may start some type of malignancy. This has been observed, for instance, in certain mouse strains genetically modified with leukemia viral genes that frequently spontaneously reactivate inducing the disease in the animals.

Oncogenes as the Seed of Selfishness

A crucial comprehension in cancer biology, gathered throughout a long period that begins with in vitro transfection experiments with oncoviruses during the 70s, has been that oncogenes and their related proto-oncogenes are often fundamental players in regulatory mechanisms of cell growth.

Cell growth and multiplication is therefore a critical point for the equilibrium of multicellular organisms but an inescapable one. In fact, the process cannot be prohibited by default in organisms, since it can be a fundamental necessity in situations like tissue repair during wounding or dead cell replacement that would impair the physiological functioning of an organ in the body. Physiological functioning also depend on a balanced tissue architecture. Consequently, cell growth must not take place for the benefit of an individual group of cells but always in an harmonic way.

Consequently, no single cell in a living tissue can be granted the autonomy to decide autonomously whether it should proliferate or remain quiescent. Cell growth is therefore regulated by the influence of neighbour cells communicating by means of growth factors (GFs) and growth-inhibitory factors; special molecules that make their way through the intercellular space, reach target cells, carrying with them specific biological messages. Positive or negative stimulations on target cells condition their proliferative behaviour.

Signalling molecules impinge on cell surface receptors, providing inputs that are transferred inside the cytoplasm by fine allosteric mechanisms involving transmembrane proteins. Inside the cytoplasm, intracellular circuits made of interacting molecules contribute in translating signals to the cell nucleus, where modulation of gene transcription determines the final answer of a cell to external stimuli.

Signalling molecules in the cytoplasm are modulated by covalent and noncovalent modifications, by binding or releasing on them specific chemical groups. The activity of signalling molecules may also be modulated by their concentration that may vary by orders of magnitude, or by their compartmentalisation in sites where they are active or inactive.

Despite similar biochemical commonalities, cell control circuits are so many and so plastic that cell behaviour in response to external signals is impossible to calculate in mathematical terms. If, for example, we consider only kinase proteins that signal through the binding of phosphate groups, there are about 500 distinct genes specifying various types of kinases. Of the about 500 kinase genes, about a hundred encode tyrosine kinases, the remainder being serine/threonine kinases. About a half of these genes make alternative splicing, encoding slightly different variant structures of kinase proteins, leading to more than 1000 distinct kinase proteins that may be present in human cells. This number provides an estimate of the active elements in just one single type of biological circuit, not to mention the almost infinite possible wirings among them.

Robert Weinberg, one of the world opinion leader in cell signalling, explain that each signalling circuit operate in a finely tuned, dynamic equilibrium, where positive and negative regulators continuously counterbalance one another. Metaphorically, a mitogenic stimulus may operate like the plucking of a fibre in one part of a spider web, which results in small reverberations at distant sites throughout the web. Clearly, neither our language nor our mathematical representations of signalling is sufficient to model, even approximately similar natural situations.

The discovery that certain oncogenes are modified versions of crucial components of cell growth regulatory circuits have revealed an underlying common strategy in cancer. It is the fact that, when their regulatory circuits are broken or sabotaged, cancer cells start acting on their own, having lost the ability to process information coming from their peers. In a word, we could say that cancer cells have become selfish.[8]

At a molecular level, this selfish growth can be realized through somatic mutations in the sequence coding a different structure of the encoded proteins. Mutations affecting the structure of oncogene products include the widest variety of possibilities: from single point mutations to chromosomal translocations of entire DNA blocks, yielding to hybrid proteins.

The biochemistry of the *Ras* oncogene is one of the best studied case involving single point mutations. It functions as a binary switch, continually flipping between active, signal-emitting and quiescent states. The mutant alleles found in cancer cells carry amino acid substitutions disabling the mechanism of shutting off. This traps Ras proteins in their 'on' state for extended periods of time, causing cell to be flooded with unrelenting streams of mitogenic signals.

The best known product of chromosomal translocation is probably that taking place in chronic myelogenous leukaemia (CML) cells. In these cells, translocation events cause the fusion of two distinct DNA sequences that unite, forming an enlarged sequence encoding a hybrid protein. One side of the translocation breakpoint is the sequences encoding the Abelson *abl* proto-oncogene; originally discovered by its involvement in Abelson leukaemia virus of mice. The resulting fusion of *abl* with a *bcr* normally unrelated sequence deregulates the Abelson protein, causing it to emit incessant growth-promoting signals. Since the discovery of the *bcr-abl* translocation, dozens of other, quite distinct translocations have been documented that result in the formation of hybrid proteins. Almost all of these are proper of hematopoietic malignancies.

However, even perfectly normal proteins can become oncogenes, when produced at abnormal levels, induced by gene amplification. In fact, oncogene such as *myc*, derive their oncogenic potential mainly from deregulation of their expression without changes in their structure. This can be due, for instance, to the integration of proviral sequences that take the control of gene expression. Proto-oncogenes can therefore be activated by insertional mutagenesis.

The avian leukaemia virus (ALV) provirus, for instance, integrates into chromosomal DNA immediately adjacent to a *myc* gene. As a consequence, *myc* expression is driven to excessive levels. Similar slowly tumorigenic viruses, such as ALV, infect cells that are permissive, releasing progeny virus particles without leaving sudden apparent sign of cell transformation. This lack of change in cell phenotypes is consistent with the fact that these viruses do not carry oncogenes.

In some human tumors, however, expression of the *myc* gene continues to be driven by its own natural transcriptional promoter, but the copy number of this gene is found to be much elevated than the usual two copies present in the human genome. For instance, in 30% of childhood neuroblastomas, a close relative of *c-myc* termed

---

[8] Additionally, one peculiar trait of cancer cells is their ability to generate their own GFs to sustain their own growth, independently from external regulatory signals sent by other cells.

N-myc has also been found to be amplified, specifically in the more aggressive tumors of this type. In both instances, the increased gene copy numbers result in corresponding increases in the level of *N-Myc* products proteins whose levels falsely signal an increase in the mitogens present in the nearby extracellular space, leading to perform incessant proliferation.

In addition, oncogenes exist that are pure viral proteins. One of the best understood of these viral oncogenes is the *tax* gene, contained in human T-cell leukaemia virus (HTLV), whose product activates transcription of two cellular genes that specify important growth-stimulating cell proteins: interleukin-2 (IL-2) and granulocyte macrophage colony-stimulating factor (GM-CSF). These particular GFs are released by virus-infected cells and proceed to stimulate the proliferation of several types of hematopoietic cells.[9]

Provirus integration, gene amplification, chromosomal translocation and other strategies all seem to converge on a common mechanistic theme. Invariably, the targeted gene is forced to work outside of its normal physiologic regulation. The net result is that a selfish behavior of cells that overgrow their natural boundaries and end up perturbing cell ecologies in multi-cellular systems.

An Additional Indication from Epidemiology: The Discovery of Tumor Suppressor Genes

As the discovery of oncogenes and proto-oncogens were being performed with experiments on cells and viruses, during the 70s and 80s, evidences of other contributions to carcinogenesis were also known from epidemiological studies, in which a sort of natural predisposition to certain type of tumors, such as retinoblastoma in children, fitted with a model in which tumor formation would take place when one hypothetical genetic entity was lost.

Alfred G. Knudson postulated the existence of a type of growth-controlling gene that, when normally present, should operate to suppress tumor formation, in other words a tumor suppressor gene (TSG). The hypothetical gene responsible for retinoblastoma susceptibility in certain children was simply termed *Rb*, from retinoblastoma. A few years after its theorisation, the first TSG was physically isolated and characterised. Today, more than 50 TSGs have been discovered and catalogued. Mechanistically there is not one single type of TSG. The only theme that unites all TSGs is the fact that their inactivation increases the likelihood that a cell will undergo neoplastic transformation.

TSGs are usually recessive, in Mendelian terms, since their loss affects cell phenotype only when both alleles are lost. In a few cases, however, haploinsufficiency

---

[9] While such proliferation, on its own, does not directly create leukaemia, populations of HTLV stimulated cells usually degenerate further, resulting into fully neoplastic cells, as it will be better described later.

can take place, since the loss of one copy of the allele is sufficient for significantly increasing the susceptibility to certain cancers. More in general, however, the familial inheritance of a single defective allele of a TSG usually results in cancer, only when the remaining healthy allele is damaged. In fact, rare inheritance of both defective alleles results in early onset of cancer in infants.

Recessivity of TSGs is confirmed by the data obtained in a famous series of experiments with chimeric cells produced by fusion of formerly normal and cancerous cells. The lack of a TSG can be restored by the presence a nucleus carrying an intact allele of a TSG. On the contrary, oncogenes from cancer nuclei are often dominant in mendelian terms.

Interestingly, it seems that incipient cancer cells struggle to become liberated from boundaries nature had imposed on them with the presence of TSGs. In fact, inactivation of one copy of a TSG is often followed by mechanisms leading to loss of heterozygosity (LOH) that incredibly facilitates the loss of the healthy allele. LOH can take place during cell mitosis and may be due to mitotic recombination, loss of a chromosomal region that harbours the gene, inappropriate chromosomal segregation, or gene conversion due to a switch in template strand during DNA replication. Interestingly, repeated LOHs have been observed in chromosomal regions that carry one or another TSG.

Rb and *p53* have been recognized as the two most important TSGs in human tumors, since they are defective in the great majority of cases, whereas almost all of the remaining TSGs are involved in the development of rare human tumors. On following we provide a succinct description of these two crucial gatekeeper genes, not only because they are almost ubiquitous role in human neoplasias, but because their loss can induce peculiar behaviours that are clearly selfish, not only in the world of cells.

The *Rb* and its gene product pRb have been defined at a molecular level in great detail. In studies that have occupied many years it has been clarified the role of the pRb protein as a crucial component of the so called 'cell cycle clock'; a crucial decision making device in the cell.

The cycle clock integrates multiple inputs coming from mitogenic factors with various other signals impinging on cell surface receptors in order to process all of them into a single decision of whether the cell should proliferate or be quiescent. It is a network of signaling proteins so complex and ordered, that it ticks a series of pre-ordained and exactly defined events during the cell life. This series of events are sorted and grouped by biologists much generically into four phases named: G1, S, G2, and M (for mitosis) that are executed always in the same exact order; giving raise to the cell cycle.

The protein product of *Rb*, in particular, is a protein that controls the passage through the so called R point during the G1 phase of the cycle, by binding or releasing the well known E2F transcription factor associated with promoters of genes involved in further progression of the cycle itself. Hypophosphorylated pRb binds E2Fs, blocking the passage through the R point, while hyperphosphorylated pRb releases them. In that way, pRb phosphorylation determines if a cell can grow or must remain quiescent.

In cancer cells, pRb function can be lost in many ways, including excessive mitogenic signals that lead to elevated levels of cyclin proteins, which physiologically control the degree to which pRb is phosphorylated, mutation of the Rb gene sequence, interference by viral or cellular oncoproteins, for instance myc, that deregulate pRb phosphorylation or activity. The net effect of pRb loss of function is that cells enter continuously into the cycle and thus keep on proliferating under conditions that would force it to halt. Accordingly, the growth-and-division cycle of cancer cells is usually not shorter than that of corresponding normal cells in the body, it is just incessantly activated. Significantly, the Rb example show us that cancer cells count more on restlessness than on speed, in their incessant race for growth.

Cancer would results from an incessant race for growth in the case of a society of cells. From a similar example, maybe we should gauge wisdom in following certain exasperated drifts in our industrialised societies.

P53, another crucial TSG in human cancers, is a subject so present in the literature that it would be futile to treat it extensively here. In brief, the impairment of p53 protein product activity induces a completely different range of situations when compared with the loss of pRb function. Its main physiological function appears to be the monitoring of the internal well-being of a cell; permitting cell proliferation and survival only if all its vital operating systems are functioning properly.

P53 protein integrates information about metabolic or genetic damage within a cell and can trigger the arrest of the growth-and-division cycle, while signalling for the repair of damage. Among others, p53 function is that of a transcription factor, imposing cell cycle arrest through its ability to induce expression of proteins that operate a cell cycle clock arrest.

Mostly its function is to emit signals that trigger apoptosis, thus inducing the elimination of a damaged cell whose continued growth might threat the organism's health. It is therefore considered a sort of self-destruction device in the cell, that is activated during cell stress, anoxia, extensive genomic damage and signaling imbalances in the intracellular growth-regulating circuits; in general when a cell is no more useful for a correct economy of an organism.

P53 inactivation is thus particularly strategic in cancer cells, since only having lost its function they can survive a variety of cell imbalances, including all the above typically present in tumor masses.

The above consideration only partially reveal how truly catastrophic is the loss of p53 function for a cell and, ultimately, for an organism. In fact, the effects of its loss continue well later in the pathology, when tumor cells are targeted with conventional chemo- and radio-therapy. These treatments are often directed toward damaging the genomes of these cells, thereby provoking their death by apoptosis. Loss of p53 function, which is seen so often in human tumors, renders them far less responsive even to therapeutic strategies.

P53 deficient cells refuse elimination, even when their conditions would strongly suggest so; in a clear cut demonstration of their selfishness. The parallelism with certain human behaviors to which we assist in our societies is sadly under our eyes everyday.

Multiple Seeds

Historically scientists have been unable to experimentally reproduce the transformation of a human somatic cell into a fully malignant cancer cell. Attempts with transforming retroviruses or DNA extracted from cancer cells in animal models seemed to indicate otherwise. In fact, in those cases the process could be realised in a straightforward way, by means of one or two crucial hits to their proto-oncogenes or TSGs.

Initial attempts to transform human somatic cells with the introduction of a single oncogene were largely unsuccessful.

Tumor viruses used in animal cell experiments carried sometimes two oncogenes that synergized in order to produce the malignant transformation. Polyomaviruses strains, like SV40 for example, bear two oncoproteins, termed *middle T* and *large T* which collaborate to transform rodent cells. Likewise, the *ras* oncogene can collaborate with the SV40 *large T* oncogene, or with a mutant *p53* gene, whereas *myc* can synergize with *middle T* oncogene or *src* to transform animal cells. Inspired by such evidences, a series of experiments of transformation were attempted on human somatic cells, by introducing sinergic couples of oncogenes. Even these attempts consistently failed to yield human cells into a tumorigenic state.

Steps ahead were made by artificially introducing the human telomerase gene (*hTERT*) in human somatic cells, taking inspiration by the fact that mice fibroblasts used in successful animal cell transformation experiments usually carry extremely long telomeric DNA and express readily detectable levels of telomerase enzyme (TERT) activity, unlikely human somatic cells which usually carry shorter telomeres and mostly lack TERT activity.

Introducing *hTERT* in addition to a *SV40 large T* oncogene (whose product inactivates both *Rb* and *p53* TSGs) highly phenotypically transformed human cell lines were finally obtained. Unexpectedly, these lines failed successive tests of tumorigenicity in vivo. In fact, when these cells were implanted into immunodeficient (actually only immunodepressed) mice of the Nude strain and resulted non-tumorigenic.[10]

The conclusion was that even TERT induced cell immortality,[11] combined with a couple of genomic hits were insufficient to generate cancer insurgence from human

---

[10] However, as it will be clearer after the reading Chapter 4, the immunity of those mice had probably had an important effect on those tumor seeding experiments.

[11] The term immortality is misleading since it, in this case, it is referred to lineage immortality, instead of an individual cell immortality, as it would be commonly intended in human terms.

somatic cells. Along these lines of research, it was demonstrated that additional hits to cell regulatory circuits were needed to finally give raise to fully malignant cells from human somatic cells, capable of growing in Nude mice. In particular, it was estimated that at least five or six mutations were required for the insurgence of human malignant cancer cells.

The essence of the so called 'multiple hits' theory of human carcinogenesis was thus delineated. The presence of oncogenes, the loss of TSGs, the intervention of immortalization, was thought to sum up progressively before a fully metastatic cancer clone could arise from a formerly normal somatic human cell.[12]

Concluding Remarks

The 'multiple hits' theory of carcinogenesis gained increasing credibility during the last twenty years of the 20th century, when it became widely accepted. The evolutionistic theory of Darwin provided the right conceptual framework for its affirmation. In turn, the 'multiple hits' theory became an important demonstration of the scientific soundness of the Darwinian perspective for evolution of life forms on earth.

Evidences contrasting this general model for carcinogenesis, such as, for instance, the fact that most often many more than five or six mutation can be retrieved in cancer genomes, or that real human cancers usually present such a genetic mess that it is impossible to pinpoint which damage is really causative from what it is only incidental, were happily put on background by mainstream science.

Historically, therefore, inconsistencies have not been fatal to this theory, and its weight for further development of cancer research has been huge and still influential even in present days. Recently, however, the advancement of novel discoveries is favoring the affirmation of more comprehensive theories of cancer insurgence which we would approach, albeit only at the end of the next chapter.

---

[12] After the advent of recent knowledge, we can anticipate at this point, despite it will be better dealt with in the next chapter, that there could have been other interpretatory ways for the resilience of human cells in transformation.

In fact, re-examining all the previous work on transfection, scientists have realised that rodent fibroblasts used in early experiments were not simply and generally "somatic" cells. In particular, they were cells with a peculiar degree of residual stemness and differentiation that had been adapted to grow in culture and propagated indefinitely.

It has recently been revealed that a single oncogene, introduced into an immortalised cell, with an unlimited replicative potential (a cell similar to a stem cell in that respect), could effectively induce the malignant transformation in an almost straightforward way.

Today, many other evidences support the view that stem cells are the crucial targets of malignant transformation. After having read the next chapter, it could be informative to step back to reexamine the early in vitro experiments discussed in this chapter. How many hits would be required to transform stem cells into a cancer clone?

Further Readings

Nature milestones in gene expression. Nat Rev Mol Cell Biol, 2005. 6:S1–Sxx

Nature milestones cancer. Nat Rev Cancer, 2006: S1-Sxxx.

Abbott, A., Cancer: The root of the problem. Nature, 2006. 442(7104): 742–743.

Al-Hajj, M. and M.F. Clarke, Self-renewal and solid tumor stem cells. Oncogene, 2004. 23(43): 7274–82.

Alonso, A., et al., Protein tyrosine phosphatases in the human genome. Cell, 2004. 117(6): 699–711.

Ames, B.N., Dietary carcinogens and anticarcinogens. Oxygen radicals and degenerative diseases. Science, 1983. 221(4617): 1256–64.

Artandi, S.E., Telomeres, telomerase, and human disease. N Engl J Med, 2006. 355(12): 1195–7.

Ashkenazi, A. and V.M. Dixit, Death receptors: signaling and modulation. Science, 1998. 281(5381): 1305–8.

Bickers, D.R. and D.R. Lowy, Carcinogenesis: a fifty-year historical perspective. J Invest Dermatol, 1989. 92(4 Suppl): 121S–131S.

Bishop, J.M., Viral oncogenes. Cell, 1985. 42(1): 23–38.

Blasco, M.A., Telomerase beyond telomeres. Nat Rev Cancer, 2002. 2(8): 627–33.

Blume-Jensen, P. and T. Hunter, Oncogenic kinase signalling. Nature, 2001. 411(6835): 355–65.

Bos, J.L., ras oncogenes in human cancer: a review. Cancer Res, 1989. 49(17): 4682–9.

Boshoff, C., Kaposi virus scores cancer coup. Nat Med, 2003. 9(3): 261–2.

Bruix, J., et al., Focus on hepatocellular carcinoma. Cancer Cell, 2004. 5(3): 215–9.

Butel, J.S., Viral carcinogenesis: revelation of molecular mechanisms and etiology of human disease. Carcinogenesis, 2000. 21(3): 405–26.

Butler, D., New fronts in an old war. Nature, 2000. 406(6797): 670–2.

Campisi, J., Senescent cells, tumor suppression, and organismal aging: good citizens, bad neighbors. Cell, 2005. 120(4): 513–22.

Comings, D.E., A general theory of carcinogenesis. Proc Natl Acad Sci U S A, 1973. 70(12): 3324–8.

Cross, M. and T.M. Dexter, Growth factors in development, transformation, and tumorigenesis. Cell, 1991. 64(2): 271–80.

Danial, N.N. and S.J. Korsmeyer, Cell death: critical control points. Cell, 2004. 116(2): 205–19.

de Caestecker, M.P., E. Piek, and A.B. Roberts, Role of transforming growth factor-beta signaling in cancer. J Natl Cancer Inst, 2000. 92(17): 1388–402.

Downward, J., Ras signalling and apoptosis. Curr Opin Genet Dev, 1998. 8(1): 49–54.

Duelli, D. and Y. Lazebnik, Cell-to-cell fusion as a link between viruses and cancer. Nat Rev Cancer, 2007. 7(12): 968–76.

Duelli, D.M., et al., A virus causes cancer by inducing massive chromosomal instability through cell fusion. Curr Biol, 2007. 17(5): 431–7.

Duesberg, P., et al., The chromosomal basis of cancer. Cell Oncol, 2005. 27(5–6): 293–318.

El-Deiry, W.S., The role of p53 in chemosensitivity and radiosensitivity. Oncogene, 2003. 22(47): 7486–95.

Evans, A.S. and N.E. Mueller, Viruses and cancer. Causal associations. Ann Epidemiol, 1990. 1(1): 71–92.

Fearon, E.R., Human cancer syndromes: clues to the origin and nature of cancer. Science, 1997. 278(5340): 1043–50.

Fidler, I.J., The pathogenesis of cancer metastasis: the 'seed and soil' hypothesis revisited. Nat Rev Cancer, 2003. 3(6): 453–8.

Fodde, R., R. Smits, and H. Clevers, APC, signal transduction and genetic instability in colorectal cancer. Nat Rev Cancer, 2001. 1(1): 55–67.

Garber, J.E. and K. Offit, Hereditary cancer predisposition syndromes. J Clin Oncol, 2005. 23(2): 276–92.

Garcia-Olmo, D.C., R. Ruiz-Piqueras, and D. Garcia-Olmo, Circulating nucleic acids in plasma and serum (CNAPS) and its relation to stem cells and cancer metastasis: state of the issue. Histol Histopathol, 2004. 19(2): 575–83.

Greenlee, R.T., et al., Cancer statistics, 2000. CA Cancer J Clin, 2000. 50(1): 7–33.

Hahn, W.C. and R.A. Weinberg, Modelling the molecular circuitry of cancer. Nat Rev Cancer, 2002. 2(5): 331–41.

Hahn, W.C. and R.A. Weinberg, Rules for making human tumor cells. N Engl J Med, 2002. 347(20): 1593–603.

Hartwell, L., Defects in a cell cycle checkpoint may be responsible for the genomic instability of cancer cells. Cell, 1992. 71(4): 543–6.

Hengartner, M.O., The biochemistry of apoptosis. Nature, 2000. 407(6805): 770–6.

Herman, J.G. and S.B. Baylin, Gene silencing in cancer in association with promoter hypermethylation. N Engl J Med, 2003. 349(21): 2042–54.

Hoeijmakers, J.H., Genome maintenance mechanisms for preventing cancer. Nature, 2001. 411(6835): 366–74.

Hunter, T., Cooperation between oncogenes. Cell, 1991. 64(2): 249–70.

Hunter, T., Protein kinases and phosphatases: the yin and yang of protein phosphorylation and signaling. Cell, 1995. 80(2): 225–36.

Hunter, T., Oncoprotein networks. Cell, 1997. 88(3): 333–46.

Hunter, T., Signaling–2000 and beyond. Cell, 2000. 100(1): 113–27.

Igney, F.H. and P.H. Krammer, Death and anti-death: tumour resistance to apoptosis. Nat Rev Cancer, 2002. 2(4): 277–88.

Kastan, M.B. and J. Bartek, Cell-cycle checkpoints and cancer. Nature, 2004. 432(7015): 316–23.

Knudson, A.G., Two genetic hits (more or less) to cancer. Nat Rev Cancer, 2001. 1(2): 157–62.

Levine, A.J., p53, the cellular gatekeeper for growth and division. Cell, 1997. 88(3): 323–31.

Liu, H., et al., New roles for the RB tumor suppressor protein. Curr Opin Genet Dev, 2004. 14(1): 55–64.

Lowy, D.R. and B.M. Willumsen, Function and regulation of ras. Annu Rev Biochem, 1993. 62: 851–91.

Luch, A., Nature and nurture – lessons from chemical carcinogenesis. Nat Rev Cancer, 2005. 5(2): 113–25.

Lutz, W., J. Leon, and M. Eilers, Contributions of Myc to tumorigenesis. Biochim Biophys Acta, 2002. 1602(1): 61–71.

Malumbres, M. and M. Barbacid, To cycle or not to cycle: a critical decision in cancer. Nat Rev Cancer, 2001. 1(3): 222–31.

Martin, G.S., The hunting of the Src. Nat Rev Mol Cell Biol, 2001. 2(6): 467–75.

Modrich, P. and R. Lahue, Mismatch repair in replication fidelity, genetic recombination, and cancer biology. Annu Rev Biochem, 1996. 65: 101–33.

Mulligan, G. and T. Jacks, The retinoblastoma gene family: cousins with overlapping interests. Trends Genet, 1998. 14(6): 223–9.

Pai, S., et al., Nasopharyngeal carcinoma-associated Epstein-Barr virus-encoded oncogene latent membrane protein 1 potentiates regulatory T-cell function. Immunol Cell Biol, 2007. 85(5): 370–7.

Parkin, D.M., Global cancer statistics in the year 2000. Lancet Oncol, 2001. 2(9): 533–43.

Parsonnet, J., Microbes and Malignancy: Infection as a Cause of Human Cancers. 1999, London, UK: Oxford University Press.

Reddy, e.p.s., a.m. curran, t., the oncogene handbook. 1998, Amsterdam: Elsevier.

Robinson, D.H., Pleomorphic mammalian tumor-derived bacteria self-organize as multicellular mammalian eukaryotic-like organisms: morphogenetic properties in vitro, possible origins, and possible roles in mammalian 'tumor ecologies'. Med Hypotheses, 2005. 64(1): 177–85.

Rowley, J.D., Chromosome translocations: dangerous liaisons revisited. Nat Rev Cancer, 2001. 1(3): 245–50.

Seton-Rogers, S., Another piece in the p53 puzzle. Nat Rev Cancer, 2007. 7(7): 488–488.

Sherr, C.J., Cancer cell cycles. Science, 1996. 274(5293): 1672–7.

Sherr, C.J. and F. McCormick, The RB and p53 pathways in cancer. Cancer Cell, 2002. 2(2): 103–12.

Steeg, P.S., Metastasis suppressors alter the signal transduction of cancer cells. Nat Rev Cancer, 2003. 3(1): 55–63.

Tapon, N., K.H. Moberg, and I.K. Hariharan, The coupling of cell growth to the cell cycle. Curr Opin Cell Biol, 2001. 13(6): 731–7.

Ullrich, A. and J. Schlessinger, Signal transduction by receptors with tyrosine kinase activity. Cell, 1990. 61(2): 203–12.

Yarden, Y. and M.X. Sliwkowski, Untangling the ErbB signalling network. Nat Rev Mol Cell Biol, 2001. 2(2): 127–37.

zur Hausen, H., Viruses in human cancers. Science, 1991. 254(5035): 1167–73.

zur Hausen, H., Proliferation-inducing viruses in non-permissive systems as possible causes of human cancers. Lancet, 2001. 357(9253): 381–4.

Chapter 3
Selfish Evolution

'In the survival of favoured individuals and races, during the constantly-recurring struggle for existence, we see a powerful and ever-acting form of selection.'
Charles Darwin. Origin of Species (1859).*

Screenshot from the initial sequence of "2001: A space Odyssey", a 1968 science fiction film directed by Stanley Kubrick. The film deals with thematic elements of human evolution, technology, artificial intelligence, and extraterrestrial life.

Introduction

.. but hardly a chance for evolution.

In the previous chapter, we have discussed how various experimental evidences piled up to sustain the carcinogenesis theory of 'multiple hits'. The model holds that holding that at least five or six crucial mutations must accumulate for transforming the genome of a human somatic cell into a fully malignant cancer cell. In physiological conditions, similar multiply mutated cell clones are incredibly improbable entities to be formed in nature.

---

* From www.brainyquote.com

M. Conti, *The Selfish Cell*,
© Springer Science+Business Media B.V. 2008

The fact that such cell clones so often arise in the human body, as confirmed by cancer incidence worldwide, has required a theoretical framework that could justify why such highly rare and deleterious mistakes of nature actually take place quite often.

The theory of Darwinian evolution offers an effective explanation for that. A modern comprehension of the Darwinian perspective holds that random mutations in the DNA enable a source of inter-individual variability among cells. Among various individuals generated in this way that compete for limited resources into a certain environment, those that happens to be the fittest, especially under the reproductive point of view, are selected and live on to the future; until a new type of environmentally fittest individual would rise and dominate the environment.

As for animals in a natural environment, in multicellular bodies, cells are thought to continuously be generated to replace those that die. The Darwinian theory would predict that with a process of natural selection, cells with genomes that enhance their survival are favoured on other cells, can multiply and soon numerically dominate their environment, i.e. their tissue of pertinence. Cells with less fit genetic traits are out-competed and doomed to die. In this way, cell clones with more and more enhanced survival traits would be selected inside multicellular systems. At the end of the process, having acquired five, six or more survival traits, cell clones so resistant and selfish would constitute malignant cancerous cells.

As it usually happens in the macroscopic world, the selection of selfish traits is exacerbated in a hostile environment. Analogously, the process of selection is sped up under the selective pressure of carcinogens. In their presence, the rate of mutation in cell genome is significantly boosted because cell death rate is increased due to their toxicity and also the inter-individual variability is increased due to their mutagenic properties. The process of insurgence of new species would therefore be greatly accelerated and the process of Darwinian selection taken to extreme.

In precancerous and cancerous tissues many factors, such as cell crowding, lack of oxygen, of nutrients, excess of waste metabolites, of toxic substances, and even cytotoxic drugs administered for cancer chemotherapy, greatly increase the selectivity of the environment and therefore increase selective pressure on cells. Aberrant cells clones are continuously selected in such microenvironment that are more and more malignant, even resistant to anti-cancer pharmaceuticals.

A similar mechanism of natural selection holds a harsh reality that first worried Darwin himself: it favours selfish individuals. The affirmation of cancer cell clones into a biological system is a dramatic empirical demonstration of that point. Therefore, the fact that cancer cell clones can arise by means of competitive natural selection has become both verification and a declaration of defeat of the Darwinian evolutionary perspective for living systems, because the prevalence of an extremely competent, super skilled and selfish killer inside them could only doom the entire system to death. It is an ultimate prediction of global collapse for living systems when single components search for their own individual affirmation.

In this chapter we will deal with the concepts of competitive evolution of cancer cell clones and how that model is being currently improved with new evidences coming from epigenetics and stem cell biology. However, we should probably also consider that competitive evolution is not the end of the story, when dealing with high order life forms constituted by the collaboration of individuals with different specificity of tasks, like human or animal societies, or like the body of multicellular organisms. The existence of similar systems in nature, which are characterised by coexistence of individual peculiarities and their integration, is an actual demonstration that evolution must have proceeded in other ways than competition, at least from a certain moment on if not from its beginning.

Intuitively, the maintenance of various peculiarities in these systems is a stronger prerogative than competition and affirmation of the stronger over the weaker. Actually, competitive escape of single components would be antithetic with the existence of complex systems, even of the simplest prokaryotic cell. Adaptability and the possibility to count on various functionalities, in a word plasticity, would appear a better prerogative to environmental adaptation. Living matter plasticity, complemented with fewer than five or six selfish traits, would be seen at the end of this chapter to be a novel, more evolved, model of carcinogenesis.

Individual Variability

DNA is the database of information instructing almost all known biological functionalities in living systems. It is considered a very stable entity within cells. Naturally occurring random mutations in the DNA sequence are caused by the effect of cosmic high energy electromagnetic radiations, of the exposure to ultraviolet components of the solar light, or because of intrinsic errors made during DNA replication by the cell apparatus itself. All these are known to be characterized by very low rates. Therefore, the probability that a cell population could advance all the way to a neoplastic state through these natural sources of mutations should be astronomically small. Rough estimates indicate that, if cancer depended on five or six hits to crucially regulatory genes, such as proto-oncogenes or tumor suppressor genes (TSGs) by means of rare mutational events, then it should never strike human populations. On the contrary, we know full well that cancer incidence in the human population is not negligible.

As already mentioned, epidemiology has revealed that intensive exposure to mutagens, correlated with many human activities, significantly raise the rate of mutation in our DNA and consequently the probability of cancer insurgence.

In addition, a wide variety of exogenous agents exist that can promote tumor formation without having a direct mutagenic effect on DNA. Among them chemicals, physical, and infectious agents whose activity is mediated by their cytotoxicity and damaging capabilities against tissues. These agents indirectly cause an excessive number of cell division cycles, which the organism is forced to perform to replace dead cells for maintaining tissue viability. Since cell duplication brings with it a certain number of miscopied sequences in DNA, the whole process results strongly mutagenic at the end, for those cells clones that are generated by such intensive rate of proliferation.

Experiments of chemically induced carcinogesis in mouse skin prove that point. Usually, protocols of mouse skin carcinogenesis are performed with highly mutagenic compounds, such benzopyrene (BP), 7,12-dimethylbenzanthracene (DMBA), or 3-methyl-cholanthrene (3-MC), applied topically on the skin. For example, after DMBA paintings, skin carcinoma would develop after several months.[1] Alternatively a single painting with DMBA is followed by repetitive painting with a second agent, tetradecanoyl-phorbol-acetate (TPA), a skin irritant prepared from croton oil that is not a mutagen but a potent stimulator of cell proliferation. After many weeks, papillomas will emerge. If TPA painting is stopped at this point, papillomas can spontaneously regress. On the contrary, if TPA painting is continued, some of these papillomas would evolve further into malignant squamous cell carcinomas. This illustrates that sustained cell proliferation, after an initial round of mutagenesis, can alone result in enough multiple mutations to induce carcinomas.

In humans, alcohol is a chemical promoter of cancers of the mouth and throat, much like TPA in mice experiments. These carcinomas are often encountered in cigarette smokers who are also consumers of alcoholic drinks. The contribution of alcohol in tumor induction seems to derive from its cytotoxic effect on the epithelial cells. In fact, cells underlying the epithelium start dividing in order to regenerate epithelial cell layers killed by ethanol. While these cells may normally divide at a low and steady rate, their mitotic work increases substantially to compensate the effect of ethanol. Being additionally and concomitantly mutated, by mutagens contained in the tobacco smoke, cell would undergo a strong acceleration in the mutational rates, leading more readily to cancer clones insurgence.

Additional evidences come from *Helicobacter pylori* induced gastric lymphomas. Seventy-five percent of these lymphomas can be cured if patients are treated with antibiotics that eradicate the bacterium. This behaviour of neoplastic lesions closely resembles that of pre-malignant lesions promoted by TPA in skin mice carcinogenesis.

Another case in point is that of hepatocellular carcinoma, strongly promoted by chronic hepatitis B virus (HBV) infections that usually continue in active forms for decades; killing hepatocytes that must be replaced by proliferation of surviving cells. The effect of the virus would therefore be that of increasing the number of cell divisions. This situation is clearly reminiscent of that induced by ethanol in alcohol consumers.[2]

---

[1] Increasing the dose of the carcinogen, results in an increase in the number of animal subjects that develop cancer later in life. Above a certain dose, almost 100% of the animals develop cancer.

[2] However, other plausible explanations are also possible. The HBV genome, for instance, does carry a weak oncogene, *HBX*, that could operate as a classic oncogene for a longer amount of time. Researchers have assessed that this oncogene is too weak, insufficient for full transformation of hepatocytes. Alternatively, HBV might act as a liver carcinogen through a mechanism of insertional mutagenesis, similarly to non-oncogene-bearing retroviruses. Extensive molecular analyses have failed to demonstrate that the genomes of virus-associated human liver cancers carry HBV genomes integrated next to critical cellular growth-controlling genes. Therefore, HBV is also unlikely to act directly as a mutagen in infected liver cells and more probably induce liver cancer through its ability to cause continuous cell proliferation, as a classic tumor promoter.

The idea is confirmed by the evidence that hepatitis C virus (HCV) is also associated with liver carcinomas. While the two viruses are totally unrelated to one another with respect to genome structure and replication cycles, they evoke very similar biological outcomes. Significantly, a variety of other types of chronic liver injuries, including that inflicted by alcoholism, are also associated with increased incidence of hepatocellular carcinomas.

During these and other chronic infection states, consequent tissue inflammation could additionally promote tumorigenesis. Indirect evidence supporting the role of inflammation in promoting cancer comes from a large number of epidemiologic observations in humans, demonstrating that anti-inflammatory drugs, both non-steroidal (NSAID) such as aspirin, sulindac, and even steroidal (dexamethasone, betametasone, cortisone and others), reduce the incidence of a variety of carcinomas in humans. People who take low doses aspirin for years experience a significantly lower incidence of various cancers compared with the rates of the same cancers in corresponding control groups.

In the mutagenetic perspective, inflammation leads to the recruitment of immunocytes that, when activated, release reactive oxygen species (ROS). These chemicals would attack the DNA of nearby epithelial cells, extremely increasing their mutational rates.[3]

All these and possibly other examples dramatically show that, despite our common perception of DNA as a very stable entity within the cell, our biological data-bank repository is a vulnerable entity. Its apparent stability reflects nothing more than a dynamic equilibrium, an ongoing battle between the forces of order and chaos. To fight the forces of chaos, human cells contain various enzymes and a variety of low-molecular-weight biochemical species that are available to confront and neutralize mutagens before they succeed in striking the human genome. Should damage be inflicted, either because mutagens have slipped through the defenses then large group of DNA repair enzymes, the so called 'caretakers', always alert to correct structural aberrations in the double helix and its nucleotides, could come into play. These enzymes are very effective in restoring the DNA to its native state.

When the DNA protective apparatus is damaged, it leads to the accumulation of mutations at a very dangerous rate. This mutator phenotype is, in fact, a common feature of many cancers. It has been realized that inherited genetic defects in the same DNA repair apparatus (in 'caretakers' genes) strongly predispose to cancer formation. Inherited defects in base-excision repair proteins, for instance, have been clearly linked to cancer susceptibility and other pathologies, such as the xeroderma pigmentosum. Inherited defects in double-strand DNA repair have been connected with breast and ovarian cancer susceptibility among patients carrying mutant *BRCA1* or *BRCA2* germ-line alleles.

---

[3] Chronic inflammation promotes the release of growth factors, as during wound healing. As we will discuss in the next chapter, the resulting recruited immune cells can contribute to carcinogenesis and metastasis in other multiple ways, in addition to the release mutagenic substances.

Additional evidences, such as premature aging syndromes, have been traced back to inherited defects in one or another component of the DNA repair system. Individuals suffering from these syndromes often show aged phenotypes already during adolescence. Consequently, cancer and aging apparently share a common root in the progressive deterioration of our genomes. The two factors can even combine, since the ageing phenotype exacerbates the reduction in 'caretaker' genes function, in detoxification, in antioxidation, in repair or telomere maintenance ability.

The general framework for the effect of age on cancer incidence is probabilistic in nature. Given a certain level of exposure to mutagens, with the time an individual would get the opportunity of acquiring five or six fatidic genetic changes in the DNA of one of their somatic cells; that would therefore become cancerous.

Epidemiologic studies have confirmed that age is a large factor contributing in the incidence of all types of cancer. If we assume a certain level of exposure to mutagens, for instance that present into an average urbanised environment, the increase in the risk of contracting a certain type of cancer is, on average, as much as 1000 times greater in a 70-year-old man than in a 10-year-old boy. Clearly, an increased level of exposure to mutagens would have the net effect of shortening the time to completion of the carcinogenic process. Formulas have been developed that predict the frequency of various cancers as a function of age. In general, for epithelial cancers as a whole, the risk of death by cancer increases approximately with the fifth or sixth power of the elapsed lifetime, thereby strongly confirming the 'multiple hits' view. Cancer is considered a degenerative disease: in other words, we observe a time dependant morphological variation of tissues from an healthy functional state toward a malignant state.

Selection of the Fittest

As already mentioned in the first chapter, histopathological examination of thousands of human biopsies collected from the epithelia of the colon have revealed a variety of tissue states, with various degree of abnormality that range from mildly deviant tissue morphologies, which are barely distinguishable from the structure of normal mucosa, to chaotic bulk of cells that form highly malignant tumors.

Scientists have managed to systematically array degenerative morphologies in the time domain, in a succession of tissue phenotypes from normal to aggressively malignant. They have imagined that such a succession depicts the course of an actual tumor development.

Scientifically, the best documented case is the evolution of colon epithelia toward colon carcinomas; because it has been relatively simple and feasible to collect samples of colon mucosa during routinely clinical colonoscopies. In certain specimens, pathologists can observe carcinomas growing directly out of precancerous (adenomatous) tissue neoformations (polyps). They imagine these outgrowths occurring routinely during the development of virtually all colon carcinomas in which rapid expansion of carcinoma clones soon overgrows and obliterates adenomatous tissues, from which they arise.

Clinical studies performed on large cohorts of patients who have undergone colonoscopy have confirmed that view. In one such study, those patients whose polyps were preventively removed experienced, in subsequent years, about an 80% reduction in the incidence of colon carcinomas. This indirectly infers that at least 80% of the colon carcinomas should derived from pre-existing, readily detectable adenomas.

Further support for the adenoma-to-carcinoma progression comes from the disease termed familial adenomatous polyposis (FAP), in which individuals inheriting mutant forms of the *APC* tumor suppressor gene (TSG) are prone to develop thousands of polyps in the intestine. The theory predicts that, with a low frequency, one or another of these polyps progress spontaneously to carcinoma. This could justify the high risk of colon cancer in these individuals.

The development of carcinomas in other organ sites throughout the body could resemble, at least in outline, the progression observed in the colon. Many other tissues, such as breast, stomach, lungs, prostate, and pancreas, also exhibit growths that can be called hyperplastic, dysplastic, and adenomatous, and these growths seem to be the benign precursors of carcinomas. However, histopathological evidence supporting tumor progression in these tissues are not so numerous as in the case of the colon. In the case of non-epithelial tissues, including components of the nervous system, the connective tissues, and the hematopoietic system, the histopathological evidences supporting tumor progression are even more fragmentary.

In general, it is possible that some of the tissue types depicted as intermediates in the classical evolutionary sequence from normal epithelial tissue to carcinoma represent dead ends rather than stepping stones to more advanced stages of the same tumor. Some tumor may even develop through a series of intermediate growths, not represented by this classic framework. However, progressive degeneration is seen by most of the scientific establishment as a hard fact that predispose to the insurgence of cancer.

Studying various genetic changes in the genomes of small colonic adenomas, mid-sized adenomas, large adenomas, and frank carcinomas the group of Peter Nowel at the Johns Hopkins Medical School in Baltimore, during the 80s, found that epithelial cell genomes in progressively malignant histotypes (during the presumed tumor progression) carry a corresponding increase in the number of genetic alterations. These changes involve both the activation of proto-oncogenes into oncogenes and the apparent inactivation of at least three distinct TSGs, this latter modification being the most widespread anomaly.

The results of the genetic analyses of human colon cancer progression has historically provided evidence that malignant clones bring with them more and more altered, oncogene rich and TSG poor, genomes. In other words, it has been demonstrated that precancerous genetic changes accompany the phenotypic evolution of cells in tissues, observable by histologists.

A series of further studies defining genetic histories of various other tumoral progressions have been attempted. A depiction in the light of the 'multiple hits' theory for bladder, pancreatic, and oesophageal carcinomas has been reported. Despite a certain number of inconsistencies present in those data, the common view that progressive genome degeneration accompanies cancer progression has been reconfirmed.

The most obvious way to rationalize this fact is that a succession of genetic changes, striking the genomes of epithelial cells, as they evolve progressively toward more malignant histotypes, is the actual driving force of carcinogenesis. For this to take place, it is necessary that among mutations that strike the genome of cells randomly, some force of selection must be in place to stabilise only those mutations that are pro-carcinogenic and oppose to those that are anti-carcinogenic.

The Darwinian perspective offers exactly that type of selective force. In fact, when a genetically heterogeneous population of cells is generated by mutational events, the selection favours the outgrowth only of those individuals, and their descendants, that happen to be endowed with mutant alleles that favour their proliferation and survival. According to the Darwinian view of cancer progression, the evolving units are individual cells competing among each other, like individual organisms competing among each other for vital resources in a restricted environment. In this view, progressively selected cell clone populations would end up dominating less favoured neighbours.

Initially, a clone would be genetically selected for a growth-advantageous mutation, like the loss of a TSG or the acquisition of an oncogene. Proliferating, this clone would span a large enough population of cells to be statistically exposed to many other mutations. Among the doubly mutated members of the population, only those with the highest growth-promoting traits would dominate the second clonal selection round. The resulting doubly mutated cells, which would survive even more effectively than their neighbours, will spawn a further new subclone that will expand and eventually dominate the local tissue microenvironment overshadowing and possibly completely obliterating the precursor population from which it arose. Quite possibly, a sequence of four to six of such clonal successions, each triggered by a specific jackpot mutation, would explain how fully malignant cancer cells end up dominating a tissue; and can even expand further to cross the border toward other tissues.

This view has historically found support in the data provided by the Nowell's group, and later of many others, that show how increasingly malignant cancer cells have an increasing number of mutations. However, given the vast number of alterations rapidly accumulating in the genomes of clinically detectable tumors, the model has been impossible to validate accurately. In particular, even when scientists are able to define the genetic history of a particular carcinoma, the succession of genomic hits or their total number and type were not always the same. In other words, it has generally been impossible to indicate which mutations are necessary and sufficient for specific stages of tumour progression, other than a few crucial 'gatekeeper' genetic lesions in well known proto-oncogenes and TSGs, in the first phases of the process.

Adapting to be the Fittest

Doubts and inconsistencies cited above have derived from a theoretical framework in which changes in the DNA sequence, that functionally modify product proteins, are normally considered to arise by a slow and gradual process that involves natural selection of DNA sequences, operating over many generations. Those doubts still remaining, with the refinement of experimental possibilities in cancer biology, a number of other processes have been discovered that have by-passed those former issues since they would enable a better comprehension of the evolution of cancer cells than the classic genetic based framework would realise. Among them, for instance, very sudden changes have been observed in the phenotype of cells, in the absence of any kind of significant change in their DNA sequence, whose explanation could be linked to the fact that it is not the DNA but its immediate surroundings that change, thereby causing a cell to activate some of its dormant capacities.

Similar phenomena happen constantly, in all metazoan, for instance, during body development. Cells with identical DNA adopt very different phenotypes, forming tissues that have apparently very little in common with each other, such as skin, brain, bones, muscles, and many others. DNA independent phenotype changes are at the origin of cell differentiation that proceed from a stem cell to all its daughter cells each specialised in peculiar functions. The exclusive focussing on the DNA code in inheritance have cast a shadow over such evidences for decades, until they came under the spotlight again with studies on chromatin and, very recently, on stem cells and cloning.

Epigenetics, the study of heritable traits that are not dependent on the primary sequence of DNA, plays crucial roles in the global shaping and maintenance of developmental patterning. This dynamic and cell type-specific changes are fundamental for the maintenance of a form of cell memory that is required for developmental stability and that often goes awry in immune disorders and in cancer. Like primary sequence of DNA, epigenetic marks are copied during cell duplication. Therefore, both genetic and epigenetic mistakes in daughter cells are correlated to the number of times a cell has divided. Epigenetic modifications, in particular, unrelated to the classic DNA based Darwinian selection, are known that are players in eliminating the activities of crucial TSGs or activating proto-oncogenes.

To provide a measure of the relative weight of genetic and epigenetic alterations in cancer, results of recent cloning experiments of cancer nuclei in mice are illuminating. They show that a mouse melanoma nucleus, after being reprogrammed into an oocyte, can give rise to an entirely viable mouse. The fact that almost all the phenotypic properties of the cells taken from cancer can be reversed by nuclear reprogramming indicates that information causing tumour is largely epigenetic. However, cloned mice generated in those experiments shows an increased incidence of melanoma in their lives, revealing that not all the nuclear modification in cancer is epigenetic and erasable during reprogramming.

Epigenetic effects on our chromosomes include global and sequence specific DNA hypomethylation, hypermethylation, chromatin protein alterations and loss of imprinting (LOI). All tumours examined so far, both benign and malignant, have shown global reduction of DNA methylation. In addition to global hypomethylation, promoters of individual genes show increased or reduced DNA methylation levels. This applies to many TSG promoters, including the well known *RB* in retinoblastoma, but also *p16* in melanoma, *VHL* in renal-cell carcinoma, *APC* and *Wnt* in colorectal cancer. Genes normally methylated at promoters but hypomethylated and activated in the corresponding tumours are *r-ras* in gastric cancer, melanoma antigen family A (*MAGE*) in melanoma, maspin in gastric cancer, *S100A4* in colon cancer, just to mention famous cases.

More generally still, it is chromatin, and not just DNA, the actual substrate for all processes that affect genes and chromosomes. In recent years, it has become increasingly evident that chromatin can impart memory to the genetics of a cell. Histone modifications survive mitosis and have been implicated in chromatin memory. Various chromatin alterations have been found in cancer. For instance, histone H3 has two universal variants, one of which, the centromere protein A (CENPA) becomes overproduced in colorectal cancer, a factor that can easily lead to aneuploidy; the most common anomaly in cancer cell chromosomes.

Imprinting is a feature of all mammals, affecting genes that regulate cell growth, behaviour, signalling, cell cycle and transport; moreover, imprinting is necessary for normal development. Loss of imprinting (LOI), as referred to activation of the normally silenced allele, or silencing of the normally active allele, of an imprinted gene, causes reduced or absent expression of a specific allele of a gene in somatic cells of the offspring in cancer clones. LOI importance in neoplasias has been shown in experiments in which embryo cells derived only from the maternal genetic complement and embryos derived only from the paternal genetic complement have been implanted in mice and have formed tumours, namely teratomas, in syngeneic mice. Documented LOI case is that of the insulin-like growth factor 2 gene (*IGF2*) that accounts for half of Wilms tumours in children. LOI of *IGF2* is also a common epigenetic variant in adults and is associated with a fivefold increased frequency of colorectal neoplasias. Other cases of LOI that have been demonstrated in cancer include *PEG1/MEST* (paternally expressed gene 1/mesoderm-specific transcript homologue) in lung cancer, *CDKN1C* in pancreatic cancer, *DIRAS3* (GTP-binding RAS-like 3) in breast cancer, and *TP73* in gastric cancer.

This short list of epigenetic alterations documented in cancer and precancerous lesions in rapidly lengthening as we write and experts have posed great expectations in this field of cancer research.

Epigenetic alterations in cancer cells, together with the fact that epigenetic modifications are at the basis of the differences between stem cells and differentiated counterparts, suggest a role of stem cells in carcinogenesis. In fact, simple somatic cells would simply have an insufficient replicative potential to span enough daughter cells to form a macroscopic tumor mass.[4]

---

[4] As first demonstrated in the early 1960s, rodent or human cells exhibit a limited number of replicative cycles in culture. The work of Leonard Hayflick showed that cells would stop growing

Like stem cells, cancer cells have long been considered immortal, in the sense that they can span an endless progeny; at least accordingly to what can be observed in vitro. In fact, many types of cancer cell are propagated in culture are able to proliferate endlessly if provided with proper conditions. Therefore, the involvement of cell immortalization in carcinogenesis has been explored since the time of in vitro cell transformation experiments, partially presented in the previous chapter. We have already discussed how, in those experiments, mice fibroblasts, easy to transform into fully cancerous cells, exhibit high levels of the enzyme telomerase (TERT) contrarily to what happens to the hard to transform human somatic cells. Accordingly, cancer cells almost always possess TERT activity, which enables elongation of telomeres, similarly, but not exactly to what happens in stem cells.[5]

It is known that, in normal tissues, stem cells are a small minority of undifferentiated cells. Stem cells appear to have an essentially unlimited ability to proliferate. Some of their progeny remains as stem cells. Their non-stem cell descendants usually enter into a state of increased differentiation with time. These cells, duplicated from stem cells, continue to proliferate only for a certain number of divisions. They are therefore termed transit- amplifying cells. At last, after many cycles, cells differentiate and enter a post-mitotic state, from which they will never re-emerge.

Experiments in the last few years have shown that cancer cells within tumor tissues are separated into distinct subclasses, much like they are in normal tissues. Neoplastic cell populations are organized in a hierarchical structure in which relatively small pools of self-renewing stem cells are able to spawn large numbers of descendant cells that have only a limited proliferative potential.

Experimental separation of tumor stem cells has taken advantage of cell surface proteins displayed by different cell subpopulations. Cells sorted by means of specific proteins on their surface have been recovered in viable form and used in biological tests, including in vivo verification of their ability to seed malignant tumors following injection into immuno-compromised mice.

In populations of acute myelogenous leukaemia (AML), for instance, cells could be divided into groups. One group represented less than 1% of the neoplastic cells were able to produce new tumors, upon injection into host immunocompromised mice. In contrast, the rest of the cancer cells were unable to seed new tumors. These

---

after an apparently predetermined number of divisions and would enter into senescence. Senescent cells remain metabolically active but seem to have lost irreversibly the ability to re-enter actively the cell cycle. Such cells will live on for months, as long as they are given adequate nutrients and growth factors, that help to sustain their viability, but they are unable to elicit the usual proliferative response observed when these factors are applied to non-senescent cells.

[5] Most likely TERT reactivation is an adaptive response in cancer cells performed to evade crisis, that form of cell consumption triggered by the erosion of telomeric chromosomal DNA ends during replication. In fact, excessive cell duplication rounds, without due elongation of DNA ends performed by TERT, would result in chromosomal shortening beyond normal limits; in consequent widespread end-to-end chromosomal fusions and karyotypic chaos. Telomeres maintenance is artificial and insufficient in cancer cells. Telomeres erosion takes place, usually resulting in marked aneuploidy.

latter cells exhibited attributes of differentiated immunocytes and a limited ability to proliferate. Subsequent experiments extended these results to human breast cancer cells prepared directly from solid tumors. Comparable results have since been obtained from brain tumor cells, lung and recently even with cells from other tissues.

Analyses of several types of chronic myelogenous leukemia (CML) have confirmed the view on cancer stem cells. The Philadelphia chromosome, which results from a reciprocal chromosomal translocation that fuses the *bcr* and *abl* genes, as already mentioned, is observed in almost all cases of this disease. A number of distinct hematopoietic cell types within CML patients carry the Philadelphia chromosome. Included are both B and T lymphocytes as well as cells of the myeloid lineage including neutrophils, granulocytes, megakaryocyte precursors of platelets, and erythrocytes. This observation provides persuasive evidence that the cell type in which the translocation origin occurs is a common progenitor of all hematopoietic cell lineages.

Highly compelling observations of stem cells' role in cancer derive also from transgenic mice in which the expression of an activated *ras* oncogene is limited to the keratinocyte stem cells in the skin. When the transgene directs expression of the *ras* oncogene in stem cells, the mice develop malignant carcinomas. In contrast, when the oncogene is expressed in the differentiating keratinocytes, benign papillomas are formed, and these tend to regress.

These various strands of evidence have converged to the view that self-renewing cells of various types are the targets of genetic changes that lead to the formation of tumors.[6]

Andrew Feinberg, at the Johns Hopkins University, has recently advanced a model in which early epigenetic changes in stem cells are considered a unifying substrate for further cancer formation. In this novel view, many of the heterogeneous properties that are commonly associated with tumour cell-growth, invasion, metastasis and resistance to therapy could be justified by different routes that altered stem cells would take along the spanning of their progeny, reacting to different microenvironmental conditions. Essentially, epigenetic disruption might perturb the normal balance between undifferentiated progenitor cells and differentiated committed cells within a given anatomical compartment, either in number or in their capacity for aberrant differentiation.

Epigenetic alterations in stem cells would set the stage for genetic differences that distinguish tumour types. Differences between tumour types might largely be due to specific genetic 'gatekeeper' alterations that arise on the background of epigenetically altered stem cells. Epigenetic abnormalities in pre-neoplastic tissues would predispose to alteration in 'gatekeeper' genes, whose normal function is to regulate 'stemness' itself. Epigenetic changes could even be additional or surrogates for genetic 'gatekeeper' alterations, such as TSGs inactivation or proto-oncogene activation.

---

[6] However, it has not yet been ruled out that committed progenitors, which normally have only a limited ability to renew themselves, may even possibly acquire unlimited self-renewal capability during the course of tumorigenesis, in ways that we do not know yet.

On a basis of stem cell epigenetic perturbation, an initiating mutation in a crucial gene (classically among the first steps in carcinogenesis, here a later albeit crucial one) within the subpopulation of epigenetically disrupted progenitor cells would initiate genomic chaos. These could be specific for tumour type; for example, mutations in *APC* or catenin encoding genes in colorectal cancer, or *BCR–ABL* rearrangements in chronic leukaemia.

Epigenetic changes might then promote chromosomal plasticity often observed in cancer cells. In fact, already more than 20 years ago, Schmid and colleagues showed that DNA hypomethylating agent 5-azacytidine caused decondensation of centromeric heterochromatin and widespread chromosomal rearrangement. Similarly, epigenetic instability in stem cells could promote widespread alterations in the chromatin leading to an unsustainable aneuploidy and a consequent genetic mess, often observed in cancer. On this background of cancer-associated epigenetic instability, the effects of mutations in oncogenes and TSGs might be exacerbated. Therefore, the risk of developing malignancy would be much higher for a given mutational event, when it occurred on the background of epigenetic disruption.

One of the most vexing problems in cancer research has been to understand how they could acquire traits that were not there previously, such as metastatic capability and MDR.[7] As discussed above, it is generally assumed that these properties are entirely due to the acquisition of new genotypes within the tumour through clonal evolution and Darwinian selection. However, it has been recently shown that explanted haematopoietic and solid tumour cells not only acquire other phenotypes in vitro or in vivo, but these phenotypes can be modified depending on the cellular and tissue matrix milieu into which they are introduced, indicating that the heterogeneity is an intrinsic and plastic property of a primary tumour.

This epigenetic disrupted progenitor model could justify metastastic properties without requiring specific mutation and clonal selection within a large tumour mass. Rather it would be an inherent property of the progenitor cell from which the tumour arises even in early stage disease, which would requires only common epigenetic (rather than rare mutational) changes. In other words, the properties of advanced tumour cells would be present all along in the initiated progenitor cells and would only require reactivation in response to the environment.

In general, the reality of stem cell's role in cancer is forcing scietists to reconsider how multistep tumor progression can occur. In a revised scheme, a small minority of cancer cells are the true tumorigenic cancer stem cells, while the rest are neoplastic transit-amplifying cells. This implies that mutations striking the cancer stem cell population can be transmitted to descendant cells in the population, while mutations

---

[7] Stem cells efficiently pump out fluorescence dye molecules, while differentiated cells do so much less efficiently. As a consequence, after exposure of cell populations to dyes, the stem cells are usually weakly stained than all others. This behavior is reminiscent of that of cancer cells. Chemotherapy resistant cancer cells are known that can perform active excretion of ionic molecules due to the actions of plasma membrane proteins, giving rise to the phenomenon of multi-drug resistance (MDR). Interestingly, the unusual high levels of MDR related proteins are increasingly expressed by malignant types of cancer cells.

that strike the far more numerous transit-amplifying cells are less influential, since these cells have only a limited proliferative potential. Consequently, the genetic evolution that had been, in some way, documented for the bulk of somatic cancer cells, referred to cells that have a lesser importance in real malignancy of a cancer. Improved models for affirmation of cancer clones are needed that must take into account the predominant and newly discovered role of stem cells in tumors.

The revised scenario has implications for the mutational mechanism that sustain tumor progression. During the formation of tumors, the mutations responsible for initiating new rounds of clonal succession should occur in cell populations that number in the few thousands at best, rather than the millions, as it were classically considered. The mutation rates required to trigger clonal successions starting from few target cells should be vastly higher than previously considered in the multistep model of carcinogenesis.

However, it is also possible that completely different ways of posing the problem should be implemented that must consider peculiar microenvironmental and functional characteristics of stem cells. In particular, the fact that these cells are particularly protected by genotoxic insults by a number of biological and biochemical strategies. The first line of protection resides in the spatial architecture of tissues. In a healthy situation, stem cells occupy inner well protected exclusive niches in tissues, and are thus sterically protected by outer electromagnetic, chemical and infectious insults. Stem cells normally can reside in these niches for an indefinite period of time and produce progeny cells while self-renewing. Many recently characterized niches have common features, including special cell–cell communications, which are organized by E-cadherin, catenin, integrin and gap-junction proteins, termed connexins that anchor the stem cell to the niche cells. But when insults are intense or persistent, these molecular architectures can go astray and stem cells, losing their normal relational background can become extremely dangerous, becoming an easy target for carcinogenesis.

Unaccounted Factors: The Role of Novel Discoveries

Data on epigenetic alterations in cancer cells are recently finding their way as important pieces into the huge puzzle of carcinogenesis. My impression is that they won't be the last. In fact, it is highly probable that some of the genetic information, still largely unclassified today, even after the completion of the human genome sequencing, could have great relevance for better understanding carcinogenesis, in a near future. Just as an example of that is the relatively recent discovery of the importance of regulatory RNA molecules, termed microRNAs (miRNAs), found encoded in the non-coding regions of eukaryotic DNA. These are 22-residues double stranded RNA molecules that can mediate the expression of target genes with complementary sequence.

In mammals, hundreds of miRNAs have now been identified, some of which are expressed in a tissue-specific and developmental stage–specific manner. For the few miRNAs whose function has been uncovered, they are important regulators of various aspects of developmental control in both plants and animals. In the few years since the inception of this regulatory RNA phenomenon, much progress has been made towards an understanding of the mechanisms by which this occurs and the identification of cellular machinery involved in RNA-mediated silencing.

Insights leading to the understanding of miRNA biogenesis may affect cancer in at least two ways. First, emerging data indicate that deregulation of miRNAs is associated with certain types of cancer. Fot instance, chronic lymphocytic leukaemia (CLL), the most common form of adult leukaemia, can be associated with a deletion at 13q14 locus in more than 50% of cases. Additionally, various other cancers (including mantle cell lymphoma, multiple myeloma, and prostate cancers) have been linked to varying degrees with 13q14 deletions.

Down-regulation of the so called miRNA-143 and miRNA-145 has been observed in colorectal neoplasias, and expression of the miRNA let-7 is frequently reduced in lung cancers, a feature that is associated with poor prognosis. Increased expression of the precursor of miR-155 has been detected in paediatric Burkitt lymphoma. Therefore, it has been speculated, based on cancer-associated alterations in miRNA expression, that miRNAs are frequently located at genomic regions involved in cancers, and their gene regulatory function may act as classic TSGs or oncogenes. Future investigations could undoubtedly reveal additional links between the mechanisms of miRNA-mediated gene silencing and cancer.

Concluding Remarks

In a colony of single cells of bacteria, researchers can watch Darwinian evolution in action. As the cells divide, mutants appear that are selected, under selective pressure that favours some mutants over others, and soon result as dominating the culture. In principle, precisely the same thing could occur within an organism. The human body typically contains about a hundred trillion cells, and many billions are shed and replaced every day. If, as it was considered in the past, cells would be produced simply by replication of other specialized cells in each tissue, our tissues would evolve, mutations would arise, and some would spread giving raise to cancer clones, as we have depicted in the first part of this chapter. The insurgence of cancerous clones in multicellular societies that are biological tissues would therefore represent a regression to previous unicellular selfishness, where clones would act independently on the global economy of the system.

However, scientists have recently been able to observe that there is a much smarter system of tissue replacement in multicellular organism that would prevent clonal escape to take place, in most cases, more than cell intrinsic mechanisms that we have mentioned in the previous chapters. In physiological conditions, epithelial tissues retain a population of undifferentiated stem cells that have the ability to grow into all the other different types of cells. When replacement is needed, some of these stem cells divide to make transient amplifying cells that further divide several times and progressively differentiate, giving rise to a number of differentiated somatic cells.

The discovery that, even in cancer tissues, cells are organised hierarchically, with the existence of cancer stem cells and a bulk of cells in intermediate states of differentiation, is greatly changing our view about the mechanisms of tumorigenesis. In fact, cancer cells could be mostly determined by stem cells which have lost regulation and relationship with their microenvironment. This would imply that acquired traits of cancer cell clones would be not acquired, but chaotically expressed in response to peculiar microenvironmental conditions. In this novel view, preservation of stem cells integrity, through protection of their microenvironmental niches, would be a factor of paramount importance in preventing carcinogenesis.

Discoveries on stem cells have provided an important contribution to recognise multicellularity, so far greatly overlooked, in cancer biology. Today, studies on cancer stem cells are rapidly filling the gap, as we are writing this book. Most probably, and hopefully, current advancements in molecular biology would lead to an even better insight into the mechanisms underlying the formation of cancer cells in a few years.

However, if we provisionally assume the model of cancer cell formation depicted obtained at the end of this chapter, still we would have not justified the affirmation of cancer as a clinically detectable pathology. In fact, in as we will discuss in the next chapters, cancer cells are not alone entities and higher level logics in our system biology should be incorporated to better understand the pathogenesis of cancer.

Further Readings

Beausejour, C.M. and J. Campisi, Ageing: balancing regeneration and cancer. Nature, 2006. 443(7110): 404–5.

Blasco, M.A., Telomeres and human disease: ageing, cancer and beyond. Nat Rev Genet, 2005. 6(8): 611–22.

Cairns, J., Mutation selection and the natural history of cancer. Nature, 1975. 255(5505): 197–200.

Campisi, J., Senescent cells, tumor suppression, and organismal aging: good citizens, bad neighbors. Cell, 2005. 120(4): 513–22.

Campisi, J. and F. d'Adda di Fagagna, Cellular senescence: when bad things happen to good cells. Nat Rev Mol Cell Biol, 2007. 8(9): 729–40.

Crespi, B. and K. Summers, Evolutionary biology of cancer. Trends Ecol Evol, 2005. 20(10): 545–52.

Dawkins, R., The selfish gene. 1989: Oxford University Press. 366.

DePinho, R.A., The age of cancer. Nature, 2000. 408(6809): 248–54.

Feldser, D.M. and C.W. Greider, Short telomeres limit tumor progression in vivo by inducing senescence. Cancer Cell, 2007. 11(5): 461–9.

Gatenby, R.A., Commentary: carcinogenesis as Darwinian evolution? Do the math! Int J Epidemiol, 2006. 35(5): 1165–7.

Greaves, M., Darwinian medicine: a case for cancer. Nat Rev Cancer, 2007. 7(3): 213–21.

Greenman, C., et al., Patterns of somatic mutation in human cancer genomes. Nature, 2007. 446(7132): 153–8.

Hartwell, L., Defects in a cell cycle checkpoint may be responsible for the genomic instability of cancer cells. Cell, 1992. 71(4): 543–6.

Khong, H.T. and N.P. Restifo, Natural selection of tumor variants in the generation of "tumor escape" phenotypes. Nat Immunol, 2002. 3(11): 999–1005.

Kinzler, K.V., B The Genetic Basis of Human Cancer. 1998, New York: McGraw-Hill.

Lengauer, C., K.W. Kinzler, and B. Vogelstein, Genetic instabilities in human cancers. Nature, 1998. 396(6712): 643–9.

Murgia, C., et al., Clonal origin and evolution of a transmissible cancer. Cell, 2006. 126(3): 477–87.

Nowell, P.C., The clonal evolution of tumor cell populations. Science, 1976. 194(4260): 23–8.

Pathak, S., et al., Telomere dynamics, aneuploidy, stem cells, and cancer (review). Int J Oncol, 2002. 20(3): 637–41.

Ponder, B.A., Cancer genetics. Nature, 2001. 411(6835): 336–41.

Vogelstein, B., et al., Genetic alterations during colorectal-tumor development. N Engl J Med, 1988. 319(9): 525–32.

Vogelstein, B.K., Kenneth W., Genetic Basis of Human Cancer, The. 2002, New York: McGraw-Hill.

Wang, E., M.C. Panelli, and F.M. Marincola, Gene profiling of immune responses against tumors. Curr Opin Immunol, 2005. 17(4): 423–7.

Armstrong, L., et al., Epigenetic Modification Is Central to Genome Reprogramming in Somatic Cell Nuclear Transfer. Stem Cells, 2006. 24(4): 805–814.

Brabletz, T., et al., Opinion: migrating cancer stem cells – an integrated concept of malignant tumour progression. Nat Rev Cancer, 2005. 5(9): 744–9.

Calin, G.A., et al., Ultraconserved regions encoding ncRNAs are altered in human leukemias and carcinomas. Cancer Cell, 2007. 12(3): 215–29.

Clevers, H., At the crossroads of inflammation and cancer. Cell, 2004. 118(6): 671–4.

Esquela-Kerscher, A. and F.J. Slack, Oncomirs – microRNAs with a role in cancer. Nat Rev Cancer, 2006. 6(4): 259–69.

Feinberg, A.P., An epigenetic approach to cancer etiology. Cancer J, 2007. 13(1): 70–4.

Feinberg, A.P., R. Ohlsson, and S. Henikoff, The epigenetic progenitor origin of human cancer. Nat Rev Genet, 2006. 7(1): 21–33.

Feinberg, A.P. and B. Tycko, The history of cancer epigenetics. Nat Rev Cancer, 2004. 4(2): 143–53.

Gurdon, J.B., J.A. Byrne, and S. Simonsson, Nuclear reprogramming and stem cell creation. Proc Natl Acad Sci U S A, 2003. 100 Suppl 1: 11819–22.

He, L., et al., A microRNA polycistron as a potential human oncogene. Nature, 2005. 435(7043): 828–33.

He, X., et al., A stem cell fusion model of carcinogenesis. J Exp Ther Oncol, 2005. 5(2): 101–9.

Herman, J.G. and S.B. Baylin, Gene silencing in cancer in association with promoter hypermethy-lation. N Engl J Med, 2003. 349(21): 2042–54.

Hochedlinger, K., et al., Reprogramming of a melanoma genome by nuclear transplantation. Genes Dev, 2004. 18(15): 1875–85.

Huntly, B.J. and D.G. Gilliland, Cancer biology: summing up cancer stem cells. Nature, 2005. 435(7046): 1169–70.

Hussain, S.P., L.J. Hofseth, and C.C. Harris, Radical causes of cancer. Nat Rev Cancer, 2003. 3(4): 276–85.

Jones, P.A. and S.B. Baylin, The fundamental role of epigenetic events in cancer. Nat Rev Genet, 2002. 3(6): 415–28.

Jones, P.A. and P.W. Laird, Cancer epigenetics comes of age. Nat Genet, 1999. 21(2): 163–7.

Marx, J., Cancer research. Mutant stem cells may seed cancer. Science, 2003. 301(5638): 1308–10.

Mathon, N.F. and A.C. Lloyd, Cell senescence and cancer. Nat Rev Cancer, 2001. 1(3): 203–13.

Mimeault, M. and S.K. Batra, Functions of tumorigenic and migrating cancer progenitor cells in cancer progression and metastasis and their therapeutic implications. Cancer Metastasis Rev, 2007. 26(1): 203–14.

Pardal, R., M.F. Clarke, and S.J. Morrison, Applying the principles of stem-cell biology to cancer. Nat Rev Cancer, 2003. 3(12): 895–902.

Sharpless, N. and R. Depinho, Cancer: Crime and punishment. Nature, 2005. 436(7051): 636–637.

Singh, S.K., et al., Identification of human brain tumour initiating cells. Nature, 2004. 432(7015): 396–401.

Chapter 4
Cult of Selfishness

> *'No servant can serve two masters. Either he will hate the one and love the other, or he will be devoted to the one and despise the other. You cannot serve both God and Money.'*
>
> *The Gospel of Luke 16:13.*

The cult of money: environmental and social disaster. Artwork by Matteo Conti

Introduction

The reductionistic approach that has pervaded science during the 20th century has greatly contributed, in conceptual terms, to the common misleading attitude among researchers of considering cancer as a cell-autonomous process. In the last few years, thanks to the improvements in histology and to better imaging techniques based on fluorescence, laser capture microdissection, molecule specific immunostaining, and others, the position of scientists has moved from a cancer cell centred primitive depiction of tumoral tissues, to a more detailed view in which a variety of different cells, in addition to cancer cells, are indeed present. These cell types often represent a small proportion of a macroscopic mass, as it arrives under the microscope of the clinician. Therefore, they were considered only tissues debris, remained after the wave of invasion of cancer cells.

Scrutinizing with new eyes cancer tissues and their microscopic structure, histopathologists have finally gone beyond the boundaries of a more or less compulsive routinely attitude to dismiss unknown features as distracting contaminants. Reviving reports of the significant presence of many types of cells in cancer tissues, in addition to the vastly preponderant neoplastic ones, have finally appeared in the literature. These are fibroblasts, myofibroblasts, endothelial cells, pericytes, smooth muscle cells, adipocytes, macrophages, lymphocytes, mast cells and others. These cells, classically depicted as bystander-normal cells, have been shown to be active elements in the process of tumor formation. Actually, these non-neoplastic cells are active essential collaborators and servants of the carcinoma cells. They actively participate in the affirmation of selfish niches growing at the expense of their own biological microenvironment, like certain bad human activities growing at the expense our precious earth environment.

The biological knowledge of such wicked tendencies in cell microenvironment is projecting the scientific research toward a completely innovative approach in oncology. The evaluation of a biological system as a whole, is a significant change in paradigm taking place in these days. An interesting observation to that regard is that the same exact genetic sequence is blueprinted in the DNA of all component cells of a human body. Today we are becoming more aware that a certain genetic program not only encodes information required for life of single cells, but contains all the information that make cells interdependent, cooperating with one another.

In this emerging global view of biological systems, we are beginning to discover the importance of respecting natural rules. This concept is not new and we all know too well what happens in human societies, when the rules regulating relationships among individuals are not equalized. In this chapter, we will observe a series of mechanisms leading to the perturbation of relationships in the microworld of cells. The results are dramatically the same both in the microcosm and the macrocosm: the insurgence of chaos and disaster, in a word cancer.

A Corrupted Environment

An healthy and ordered environment, both intended in a social and an ecological sense, is a vital constituent of life on hearth. Even microscopically, order is essential for the life of a biological system. Tumorigenic potential of cancer cells, built up by a variety of factors, some of which we have discussed in the previous chapters, can be realized only when cells find themselves into an environment that they could subvert and colonise. In other words, cancer cells can form clinically detectable tumors only when they are able to breach the biological texture in which they reside.

An ordered, physiological tissue architecture is a potent deterrent to tumor development, even if cancer cells are actively expressing transforming oncogenes.[1] The analogy with human society could easily be envisioned since order and respect are the best deterrent to various forms of delinquency.

The dependence of carcinogenesis on the status of tissue microenvironment integrity was first documented in the 70s, when Harold Dvorak and colleagues showed that wounding was one important factor influencing tumorigenesis. In their experiments, when Rous Sarcoma Virus (RSV) was injected into a chick wing to produce a local tumour, a second tumour would only develop if a wound was simultaneously inflicted at a remote site.

The connection between cancer and wounding is long known and referred to by the old way of saying that tumors are 'wounds that do not heal'. It appears as if the healing process is continuously started but can never reach its completion in cancer tissues.

Epidemiological data indicate also that many human cancers arise in association with chronic inflammation, due often to bacterial or viral infection, chronic diseases, chemical or physical stressors. Disruption of the microenvironment by inflammation can take place for a series of factors, most of which combines in precancerous lesions. Inflammation and wound healing are intertwined processes. They often sinergistically lead to a disordered microenvironment, where host cells continuously fight parasites or toxic substances, usually being unable to resolve the situation. In these areas, there are plenty of opportunities for cancer cell clones to rise, and to subjugate the otherwise present biological equilibria[2].

In addition to disrupting the microenvironmental texture, inflammation and wounding promote tumorigenesis because during these physiological states, a series of cell signalling mechanism takes place that are hijacked by cancer cells to their own

---

[1] In order to do so, they must be able to finely sense their tridimensional microenvironment. Microenvironmental structure is sensed by cells through integrins, proteins specialised in touching and sensing the surrounding of a cell, thereby mediating adaptive responses of the cell. Mina Bissell and colleagues showed, during the 90s, that blocking integrins function was sufficient to revert the malignant phenotype of human breast cancer cells both in culture and in vivo.

[2] An escape from a similar desperate situation could be realised in many ways, typically by the formation of fibromatous masses to limit the contact with insulting agents without insisting in the fight. This is often not possible when the underlying biological order is compromised to the point that physiological cell interactions can no longer be correctly established.

advantage. This concept is clarified if we observe what happens during the process of physiological wound healing.

Immediately after wounding, vasoactive factors are mechanically released from the site of injury. These factors are signalling molecules that readily increase the permeability of blood vessels near the injured site, enabling the permeation of macromolecules from blood plasma. They constitute a recall for other cells specialised in the repairing process. Blood platelets aggregate and release active molecules, among which, predominantly, platelet-derived growth factor (PDGF). The PDGF released by platelets further attracts fibroblasts and stimulates their proliferation, activates them to release matrix metalloproteinase (MMP) enzymes.

MMPs begin to degrade the extracellular matrix (ECM), allowing structural re-modelling of tissues, carving out space to accommodate new cells. Secondarily to the degradation of the ECM, a variety of other growth factors (GFs), previously tethered in inactive forms to the ECM, are released. This additional stream of trophic signals attract immunocytes: monocytes, macrophages, neutrophils, eosinophils, mast cells, and lymphocytes. These immunocytes usually scavenge and remove foreign matter, bacteria, and tissue debris from the wound site. Concomittantly, the release of mitogenic factors, such as the well known vascular endothelial growth factor (VEGF), proceed to stimulate endothelial cells in the vicinity to multiply and to construct new capillaries, in a process known as neoangiogenesis, which is of paramount importance for cancer growth, as we will better define later.

With neoangiogenesis we can observe an interesting example of interaction between the genome and the environment. The genome is considered to determine the overall layout of the vessels in neoangiogenesis. In fact, during embryonic development, the overall size and architecture of vasculature appears genetically programmed, as it is similar between progenitors and progeny. During wound healing, however, the routing of individual capillaries appears more determined by contingency and heterotypic interactions operating over short distances, resulting in a disordered and mulfunctional meshwork.

The neoangiogenetic process, when abruptly activated, generates immature capillaries, with largely incomplete fenestrated walls that allow the passage of macromolecules, in particular fibrinogen, that are accessory for normal tissues repair. Accumulation of clumps of fibrin is evident in wounded sites and it is often present in tumor specimens.

Also epithelial cells, which are important actors around the edges of a wound, in order to reconstitute the epithelial sheet that existed previously, show cell behaviors which strongly relate wound healing with carcinogenesis. These cells reduce their adhesion to the ECM, especially to the basement membrane that separates them from the stromal compartment. By severing protein molecules of E-cadherin that bind them together, they increase mobility. While modifying their display of cell-surface cadherins, epithelial cells undergo a major change in phenotype, which causes them to assume a fibroblast-like appearance. This profound phenotypic shift, termed epithelial-mesenchymal transition (EMT), has already been mentioned in the first chapter, because it enables the epithelial cells to become motile and invasive in metastasis.

The traits acquired during the EMT allow cells to move and they fill in the gaps created by the wounding. The EMT, like other phenotypic traits is only temporary. After the reconstruction of the epithelium, cells revert to an epithelial state via an inverse program, the mesenchymal-epithelial transition (MET). As a consequence, once wound healing is complete, cells in the reconstituted epithelium show no trace of having transiently lived a mesenchymal life.

Wound healing and carcinogenesis share additional biological mechanisms, in addition to the example just mentioned. Recent observations have confirmed that even gene expression patterns in the tumor stroma are similar to those in stroma during wound healing. Analyses of a large group of human breast cancer cases, for mutations in the *PTEN* and *P53* tumor suppressor genes, have shown the presence of somatically mutated alleles of these genes in stromal cells, mainly fibroblasts, isolated from the tumors. In other tumors, distinct mutant *P53* alleles have been found also in the stromal compartment. These observations have opened the door to the possibility that genetic evolution of neoplastic epithelial cells can even be accompanied by changes in the genomes of nearby stromal cells.

Criminal Conspiracies

From the examples just summarily depicted above, we have been introduced to the world of cell communications that regulate biological processes through an amazing exchange of signals consisting of molecular interactions, involving GFs, chemokines and hormones.

In societies, the role of communication is of paramount importance for determining the behaviour of individuals. When information is corrupted it generates mistaken behaviours. This is why information is sabotaged by those who want to promote perverse logics in societies. It appears as if altered communications among cells is one of the most important factors for the affirmation of criminal associations that are tumors. On following, we provide some example of altered signalling among different cell types that strongly promote the insurgence of cancer.

The best known example of altered communication between cells of different type, leading to altered behaviour of cells, is that taking place between fibroblasts a carcinoma cells. In a physiological situation, fibroblasts play an important role in preventing the progression of transformed epithelial cells. In a well known experiment, it was demonstrated that normal fibroblasts cultured together with prostatic carcinoma cells could promote the reversion of neoplastic cells to a non-tumoral phenotype. The role of transforming growth factor β (TGF-β) supplied by fibroblast is fundamental for suppressing epithelial transformation, mainly controlling mitogenic signalling in the adjacent tumor cells via paracrine mechanisms involving other GFs.

Experiments have been performed with genetically engineered mice, in which germ line had been mutated in order to inactivate the TGF-β receptor. In those models, stromal cells are no longer susceptible to TGF-β inhibitory effect, and stromal hyperplasia occurs in many of their tissues. Hyperproliferating fibroblasts can drive nearby epithelial cell layers to proliferate and, eventually, to develop into carcinomas; demonstrating the power of stromal cells to stimulate epithelial cells proliferation, doing so in ways that can lead to neoplastic transformation.

Many studies have shown that fibroblasts resident in established tumors differ considerably from those in normal tissues. They are atypically activated, displaying markers of smooth muscle cells phenotype. In tumors, these cells are termed my-ofibroblasts, or carcinoma-associated fibroblasts (CAFs), and have been shown to enhance malignant epithelial transformation. Thus, stromal fibroblasts can have a bipolar role in cancer, depending on their phenotype. A myofibroblast-rich stroma represents a force that can drive aggressive tumors affirmation.

The collaboration of CAFs with cancer cells has been confirmed in experiments in which mixtures of normal fibroblasts and cancer cells or CAFs and cancer cells have been injected in mice. The grafts containing CAFs plus cancer cells formed tumors much more efficiently than mixture containing normal fibroblasts. When injected alone, the CAFs formed no tumors at all.

Alleys, such as CAFs, can confer multiple additional functionalities to nearby epithelial cancer cells. Possibly the most important of these benefits is their contri-bution to the realisation of neoangiogenesis, a crucial step for further tumor develop-ment. In fact, like normal cells, but more frantically, tumor cells require access to the circulation in order to receive nutrients, oxygen and other supplies for their growth.

Early angiogenetic areas are initially circumscribed, then intense angiogenesis can begin when cancer cells start to become invasive, penetrate the basement mem-brane, and acquire direct, intimate contact with stromal cells. This sudden, dramatic change in the behaviour of tumor masses is termed the 'angiogenic switch'. Tumor invasiveness and intense angiogenesis are often tightly coupled processes.

Within a tumor mass, long after the switching of the angiogenesis, the weakly angiogenic cancer cells rely on help from their 'normal' foes, stromal cells, in order to proceed in further neoangiogenesis. In this phase, the tumor-associated blood vessels may sprout directly from existing vessels in adjacent normal tissues. It is likely that the endothelial cells forming these new capillaries arise through the pro-liferation of endothelial precursor cells (EPCs), residing in already-formed vessels.

Because angiogenesis is usually a rate-limiting step in tumor formation, the tumor-stimulating effects of CAFs may be largely due to their ability to recruit new EPCs. CAFs can release chemokines that serve to recruit additional circulating EPCs into the tumor stroma. Prominent among them is VEGF, whose production is realised by CAFs, but also by tumor cells and macrophages, depending on the tumor type.

VEGF released into the circulation stimulates the release of EPCs in the bone marrow and guide them to the tumor mass. Once they arrive to destination, EPCs are induced to differentiate into functional endothelial cells that construct the tumor-associated vasculature. These dynamics of EPCs recruitment dramatically illustrate that primary tumors can perform long-range interactions throughout the body, even before leaving the primary site to colonise distant tissues.

The formation of blood capillaries involves a number of other factors, in addi-tion to the VEGF. TGF-b, basic fibroblast growth factors (bFGFs), interleukin-8 (IL-8), angiopoietin, angiogenin, PDGF and many others give different contribution to angiogenesis, but under the same finalistic premises and similar mechanisms of realisation. This variety of trophic factors parallel that several distinct cell types, in addition to endothelial cells, normally take part in the construction of capillaries and larger vessels.

Unbalanced heterotypic interactions taking place during tumor formation, determined by the presence of wrong amounts of those signalling substances result in the formation of capillaries that are different from those of healthy tissues. The overall meshwork of blood vessels around and within tumor masses is chaotic and, at a microscopic level, the capillaries show too many gaps in their walls. They are defined for that 'leaky capillaries' that are responsible for the deposition of fibrin in the tumor parenchyma as well as for enabling free access to many other molecules from the circulation. Quantitative measures indicate that the walls of capillaries in tumors are about ten times more permeable than those of normal capillaries.

The leakiness of these capillaries leads to the extravasation of excessive amount of fluids in the parenchymal spaces within a tumor. In normal tissues, these fluids are drained by lymphatic vessels, which eventually empty their contents into the venous circulation. Within solid tumors, instead, the ongoing expansion of cancer cells exerts abnormal pressure on lymphatic vessels causing their collapse and malfunctioning. The resulting defective lymphatic drainage within the cores of solid tumors further exacerbates the elevated accumulation of fluid caused by capillary leakage, generating relatively high fluid pressure in the tumor. This pressure constitutes one of the major forces that oppose to the penetration of anti-cancer therapeutic drugs within solid tumor parenchimas, and is considered one of the main determinants of the failure of chemotherapeutic regimens currently adopted.

Due to the paramount importance of neoangiogenesis in tumor development, the idea of strangling tumors by cutting off their blood supply was pioneeristically proposed by J. Folkmann, as early as in the late 70s. Successful therapeutic strategies based on neoangiogenesis inhibition were developed in mice. Unfortunately, the success in mice experiments could not be repeated in humans. However this line of research keeps on being actively pursued by many researchers and pharmaceutical companies; even because it has been acknowledged that anti-angiogenetic effects play a greater role in conventional anti-tumor therapies than anyone had imagined. Clinical responses to certain types of conventional chemotherapy are strongly influenced by the sensitivity of the tumor-associated microvasculature to these agents. This suggests that in the future, treatments with many anti-cancer chemotherapeutics may be optimized by gauging their effects on the tumor-associated microvessels rather than on the tumor cells themselves. Quite recently, novel anti-angiogenic molecules (in particular anti-VEGF antibodies) have found their way in the clinic and are at the basis of modern cancer therapeutic protocols.

There could be many more examples to illustrate deranged interaction among cells, in addition to those above mentioned in the case of neoangiogenesis. We would just add only one more example, because we hinted to it in the first chapter. It is the role of heterotypic interactions in metastasis. It is particularly important for it gives a justification at a molecular level of the Paget's 'seed and soil' metaphor.

Let's take a well studied case: bone metastasis of breast cancer cells. Bone metastasis is a frequent event, even if most types of cancer cells are, on their own, would be incapable of remodelling bone structure. To colonise the bone, cancer cells must therefore recruit osteoclasts and osteoblasts, that normally would work in coordination among each other in response to remote hormonal impulses and growth factor signals, in order to maintain bone homeostasis.

Cancer cells reach the bone through vessels feeding the marrow, probably attracted by unusually rich sources of mitogenic and trophic factors in the bone matrix. Upon arrival, breast cancer cells revert to a phenotype characteristic of their precursors milk-producing epithelial cells (MECs), which are able to release hormones that trigger osteoclasts to dissolve the bone matrix. In fact, during lactation they release in the blood the same hormones that are useful for increasing the availability of circulating calcium used in the mammary gland. Osteoblasts, stimulated to dissolve the bone matrix, liberate rich supply of proteins formerly attached to it. Liberated GFs fuel further growth of breast cancer cells, inducing them to secrete more hormone molecules, thus triggering a 'vicious cycle'.

With different strategies, but with similar goals, other carcinoma cells can activate osteoblasts rather than osteoclasts, producing osteoclastic lesions. This is the case, for instance, of metastatic prostate cancer cells that, once in the bone marrow, produce endothelins that acts via heterotypic signalling to stimulate osteoblasts.

In general, the adaptability of cancer cell phenotypes to specific tissues and thus the ability to acquire suitable phenotypes, to succeed in persuading invaded tissue cells to collaborate, may depend on the degree of plasticity of their epigenomes, at the time of their arrival at the site of colonisation. This would lead us back to the role of stemness for cancer cells.

The existence of similar heterotypic interactions reveal a degree of complexity just a few years ago unimagined in cancer research. However, many additional level of complexity are rising on the horizon of cancer research. One of them is constituted by the interaction of the immune system with cancer cells, which is the issue of the next session. Many other systemic interactions are probably equally important during carcinogenesis, like those with the nervous and neuroendocrine system. The study of these aspects would hopefully contribute to explain in the future the increased predisposition to cancer registered in association with depression, anxiety and stress. These are not the subject of this book, but appear promising issues for future research.

War Zones

In wounds and, more in general, in tissues contaminated by carcinogens, cancer cells can arise, subvert physiological interactions and find alleys in stromal cells to supplement their ascent to power. However, the raise of dangerous elements among cell societies is not as passively observed as it was thought in the past. Complex interaction between components of the immune system and rising cancerous clones are constantly happening that strongly determine if a cancer would ever reach clinical interesting size.

In a schematic metaphoric view, the immune system is an army composed of various types of soldiers with very specialized tasks. Macrophages and natural killer (NK) cells, for instance, have an innate ability to recognize enemies that should be destroyed. These cells are therefore the main actors of the so called innate immune response. Other soldier cells are T lymphocytes (CTLs), whereas helper T lymphocytes (THs) and regulatory T cells (Tregs) are like sentinels that recognize antigenic targets and instruct others to perform the kill. These cells are components of the so called adaptive arm of immunity.

Normal cells throughout the body routinely present oligopeptide fragments on their surface, elaborated from degradation of intracellular proteins, mounted on membrane major hystocompatibility complex class I (MHC1) proteins. These epitopes are scrutinized, sensed by immunocytes to check for the presence of altered proteins that indicated the presence of infectious agents, mutations or other possible anomalies. Similarly, professional antigen-presenting cells, such as macrophages and dendritic cells use MHC class II receptors to present oligopeptide fragments of proteins that they have scavenged from microenvironment and processed intracellularly.

All recognition mechanisms known so far are based, metaphorically, on tactile sensations, because they involve glycoprotein interactions determined by shape complementarity. Specificity of binding of molecular sensing devices of immunity is modulated by a process of gene segment rearrangements in the genomes of immunocytes that consequently modulate shape of proteins and specificity of their targets. It is well known, for instance, the case of genetic scrambling responsible for the modulation of binding properties of soluble antibodies,[3] that recognize and bind complementary molecular structures both on the surface of cells or freely diffusing in solution. The binding of antibodies marks their targets call for the killing of macrophages, NK cells or the complement system.

One of the major questions in cancer research, during the second half of the past century, has been whether the immune system could recognize cancer cells and, in the affirmative case, if it can eliminate them. Today we know that the immune system does indeed contribute to protect the body from tumor cells. When complete remissions are obtained, evidences have been collected that the immune system plays a substantial role in the process of complete eradication of cancer cells. Too often, however, cancer can prevail.[4]

A series of experiments, performed already in the 50s, showed that the force of the immune system can be enough to eliminate a tumor, when the right components of immunity are involved. In these experiments, tumor pieces removed from mice were implanted into mice of the same strain. These tumor fragments were rapidly

---

[3] Increased cancer susceptibility is registered in genetically altered mice lacking *RAG-1* or *RAG-2* genes that are responsible for the rearrangement of soluble antibodies, as well as antigen-recognizing T-cell receptors.

[4] During cancer progression, especially in advanced state of the pathology, the immune system is very weakened, even because of the immunotoxic side effects of chemotherapy that patients can die of parasitic bacterial infections.

destroyed by the host immune response. Tissues from one strain of mice were recognized as foreign and eliminated, on the basis of a typical allogeneic response, as today we know it happens, for instance, in organ transplantation among incompatible individuals.[5]

The immune system should therefore vigilate and counterfight tumors, when it is able to recognise them. However, the phenomenon of immunosurveillance, has even suffered major drawbacks in the past, in particular after it was shown that immuno-compromised Nude mice (hairless mice lacking a functional thymus so that T lymphocytes development is impaired) were as susceptible to spontaneously arising or chemically induced tumors as fully immuno-competent mice. Only years later, it was shown that Nude mice, while lacking normal T lymphocytes retain important components of their immune system in intact forms. In particular, they possess NKs in large numbers that are very active against cancer cells.

Among the difficulties encountered by the immune system in establishing an effective anti-cancer defense is paradoxically one of its most important prerogatives: self-tolerance. The immune system is in fact able to recognize and leave the body's own tissues intact. Cancer cells, that are native to the body, could constitute a target difficult to recognize. Scientists believed, especially in the past, cancer cells should be transparent to the eye of immunity.

Intuitively, however, mutated genes in cancer cells can produce mutated proteins that are recognizable by immune cells. In fact, the common view was that cancer tissues are unrecognizable by immunity was proved wrong, already during the 80s, when it was shown that chemically induced tumors in mice were antigenic. Some of the tumor antigens are of the type that attracts the attention of the immune system, because they are clearly not self to the body. An obvious example of a specific cancer antigen is provided by the ras oncoprotein, which are created by amino acid substitutions in normal ras proteins. Similarly, *p53* mutants carry specific amino acid substitutions that might cause the altered versions of this protein to be immunogenic. Another example is the bcr-abl fusion protein, in chronic myelogenous leukemia (CML) cells. Similarly, other chromosome-fusion derived proteins in other malignancies are immunogenic.

In addition, tumor associated antigens (TAAs) exist that are normal epitopes, but expressed at abnormal levels compared with corresponding physiologic situations. The catalytic subunit of TERT , for instance, not expressed at detectable levels in somatic human cells, is expressed at significant levels in human tumor cells and could therefore by recognized by the immune system. Human melanomas, in another example, overexpress a class of cell surface carbohydrates that can also provoke an immune response. In fact, expression of one of these gangliosides, termed

---

[5] In certain cases, however, cancer can spread among allogeneic individuals even in the presence of an efficient immune system. This is the case, for instance, of canine cancer spreading among wild dogs. These tumors are transmitted through wounds and bites inflicted during fights among the animals. These fortunately rare events of cancer cell transmission among allogeneic animals indicate that, when cancer cells are particularly virulent, they can succeed in overcoming the immune system and in establishing macroscopic tumors, despite immunosurveillance.

GD3, is sometimes higher in melanomas than in normal melanocytes. In each of these cases, the over-expressed proteins or carbohydrate moieties can attract the attentions of the immune system. These are associated with the induction of autoimmune reactions, in patients in which the immune system attach these targets. Embryonic proteins, when expressed in adult tissues, might also represent an additional source of these TAAs, whose expression is not limited to malignant tissues.

The fact that immunosurveillance exist, but cancer cells can manage to evade it and establish macroscopic tumors, indicates that they are able to find ways of escaping elimination by immunity. In fact, tumor cells have been shown to possess a series of prerogatives that can help them evading safety controls, launching counteroffensives against immunocytes, and even enslaving them to their own side of the battlefront. Scientific evidences for these maneuvers have been gathered only very recently. Current hope among scientists is to learn how to support the immune system to fighting back and win.

Most of the prerogatives of cancer cells are thought to derive from the phenomenon of immunoediting, a form of genetic adaptation of cancer clones in response to the attacks of immunocytes. For instance, it has been demonstrated that when chemically transformed cells arise in immunocompetent mice, cells that are strongly immunogenic are effectively eliminated by the host immunity. Weakly immunogenic cancer cells, on the contrary, can survive and outgrowth the others. In other words, the immune system exerts a selective pressure on cancer clones, leading to the selection of the fittest (less immunogenic) ones.

Immunoediting could be envisioned as a particular type of Darwinian selection, in which the selective force is constituted by the attack of the immune system on incipient tumor cells. Once again, the selection of the fittest is proved a deleterious process in the survival of a biological system, as a whole, since it leads to the insurgence of evolved selfish subjects that threaten the existence of the system as a whole. A similar conclusion was reached when discussing Darwinian evolution of cancer clones under selective pressure of cell killing factors of totally different origin.

War Maneuvers

Clinical observations started during the 90s have provided indications on the pivotal role of the immune system in the insurgence of cancer in humans. Increased susceptibility to specific types of cancer has been documented in patients that had received organ transplants and whose immune system was controlled by long-term treatment with immunosuppressive drugs.

The most common cancers among transplanted patients are those associated with viral infections, including Kaposi's sarcoma, caused by HHV-8; lymphoproliferative diseases, including the Hodgkin's disease mostly associated with Epstein-Barr virus (EBV); squamous cell vulvar and anal carcinomas, triggered by HPV; hepatocellular carcinomas, caused by HBV or HCV; and uterine carcinomas, also associated with HPV infections. Similarly, increased rated of virus-induced malignancies are seen

in patients who are congenitally immunodeficient or have acquired the immunode-
ficiency syndrome (AIDS).

Most of the populations of the western world are infected by those types of
viruses with an incidence much higher than that of corresponding malignancies.
This fact support the view that viruses are not the sole cause of malignancy and
those other factors should come into play. An extremely weakened immune system,
for instance, is considered strongly predisposing for cancer viruses attack. However,
the existence of peculiarly carcinogenic variants of common viral families affecting
humans have not been described in the literature nor thoroughly ruled out, thus far.

Whether a competent immune system also erects defenses against the great majority of human tumors, of non-viral origin, is a different matter. A definitive answer to this question has not yet been given. In general, however, whatever tumor
we consider, there are traces of past and ongoing battles between cancer cells and
immunocytes, because we can find weapons on the battlefield. In fact, a number
of research reports have demonstrated the presence of anti-tumor antibodies in the
blood of patients suffering from various types of cancer, even of non viral origin.
These are most commonly antibodies against cancer epitopes. Apparently, however,
antibodies are not sufficient or efficient in the elimination of tumors from the organism. This is probably due to the fact that cancer cells are not passive targets for
elimination.

One of the main strategies used by cancer cells is modulating in the expression of
antigens that elicit the attention of the immune system. Shedding proteins could, in
principle, compromise cell functions related to survival and proliferation. However,
when antigen-negative variant cells manage to survive, they may even be able to
escape immune recognition and eventually emerge as the dominant cell population
in a tumor mass, while other cells are eliminated by immunity.

In many tumors, cancer cells cannot simply repress expression of an antigen,
because the expression of a particular oncogene product may be essential for their
existence, as in the well known case of the HER2/Neu protein in breast cancer cells.
This protein well stimulates an immune attack; however, the neoplastic cells in
breast carcinomas cannot afford to shut it down, because of its critical for proliferation and escape from apoptosis.

In similar cases, neoplastic cells may resort to alternative strategies to avoid
immune recognition, such as repressing the expression of MHCI receptors that
display intracellular antigens on their surface. Many types of human cancer cells
have been found to lack normal levels of MHCI molecules, thereby preventing potentially antigenic oligopeptides from appearing on their surfaces. Loss of MHCI
protein expression is often associated with more invasive and metastatic tumors.
For example, in more than half of advanced breast cancers, the display of these
critical antigen-presenting molecules has been totally lost, and such loss correlates
with poor prognosis. Carcinoma cells that form micrometastases in the bone marrow
often show little if any MHCI expression, probably in response to intensive surveil-
lance by the immune cells operating in this tissue environment.

This mechanism is closely connected with the role of interferons (INFs), in particular the effect of IFN-γ is connected to the activity of NK cells. Once NK cells

identify targets for elimination, they release IFN-γ in the surrounding that interrogates potential target cells. In fact, IFN-γ stimulates cells to display on their surface increased levels of MHCI molecules, in a process that resembles a request of credentials. Should they result guilty of carrying unauthorized antigens they would be attacked and probably destroyed by NKs. Simultaneously INF-γ enables the recruitment of other types of immunocytes, to assist in the eventual killing. Total absence of MHCI molecules, as well, would call for an attack by NK cells, which continuously patrol the body's tissues looking for cells that have lost these credentials from their surface. This explains why researchers have found that tumor cells can even selectively suppress the expression of a few specific MHCI molecules. This type of selective suppression may block the specific presentation of particular tumor antigens, allowing the tumor cell to survive free of attack by immunocytes.

NK cells can additionally recognize cancer cells in other ways, for instance sensing the presence of one or more stress-associated proteins, like for instance well known heat shock proteins (HSPs) on their surfaces. Therefore, metastasizing cancer cells, especially when travelling into the blood circulation, where they are particularly exposed to immune attacks, have adopted a strategy of covering underneath a cloak of adhering platelets, preventing immunocytes from reaching them.

Also normal cells protect themselves from inadvertent killing by immunity. For example, they may express anti-complement proteins on their plasma membranes, namely membrane-bound complement regulatory proteins. The most important of these are CD46, CD55, and CD59 that have interestingly been found to be overexpressed on the surfaces of a variety of human cancer cell types. Similar overexpression could afford cancer cells a measure of protection from complement-dependent killing. Interestingly, analogous strategies are adopted by certain viruses. Herpes virus saimiri, a virulent herpes virus that causes lymphomas and leukaemias in monkeys, expresses proteins closely related to human CD59, which apparently protect virus-infected cells from rapid complement activity. Several other mechanisms that protect cancer cells against complement have been mentioned in the literature but remain poorly studied.

In addition to these shedding strategies, cancer cells can actively counter fight immunity. For instance, they have been observed to kill immunocytes by turning their weapons, such as the Fas-Fas ligand apoptotic inducing machinery, against themselves.[6]

Alternative forms of counterattack involve the release of interleukin-10 (IL-10) or TGF-β. Both of these secreted proteins are potently immunosuppressive and act through their ability to induce T lymphocytes to enter apoptosis. In addition, TGF-β can induce apoptosis of dendritic cells and macrophages, the two key antigen-presenting cells of the immune system. The release of IL-10 by human cancer cells is mimicked by EBV, which has acquired and remodeled the cellular IL-10 encoding

---

[6] Binding to Fas-receptor by Fas-ligand carried by lymphocytes can trigger apoptosis in targeted cells; however certain cancer cells can release soluble Fas-ligand like molecules that go activating the Fas-receptor system on immunocytes.

gene. As we have already mentioned, this virus is an important etiologic agent of Burkitt's lymphomas, other lymphomas of the B-cell lineage, and nasopharyngeal carcinomas. By forcing virus-infected cells to release an IL-10-like cytokine, EBV protects these cells from direct attack by cytotoxic immune cells. Similar observations points in the direction that the existence of peculiarly virulent strains of EBV, for instance, could exist that are particularly virulent or even carcinogenic.

In general, these and many other battlefield maneuvers on the front of cancer cells are accounted for on the basis of the immunoediting theory. However, the extreme plasticity, speed and variety of strategies employed, hardly fit properly into the model of clonal selection. These properties are simply too rapidly acquired and it appears as they can just be switched on and off at the right place and time. A more dynamic scenario is that in which antigenic cancers, suffering the attack by one or another arm of the immune system, find ways to adapt their antigen expression phenotypically. Similarly, counteroffensive and hiding maneuvers are not developed through the selection of rare mutant cancer cells, but by the activation of functionalities already present in their genome. The variety of strategies strongly implicates a certain degree of pluripotency in cancer cells, once again underlining the role of cell stemness in carcinogenesis.

Allied and Spies

The open battlefield scenario depicted above, is currently being questioned by novel evidences revealing more subtle manuevers among cells. In fact, in real situations, it appears that covert strategies of persuasion, based on molecular cross-talks among cells, could be an important element in tilting the balance toward cancer affirmation. We are just beginning to realize that in tumors immunocytes, such as macrophages and regulatory T lymphocytes (Tregs), that are normally responsible for self-tissue recognition and tolerance, can be hijacked for providing immunotolerance to cancer cells. These evidences are widening our perspective on how deranged heterotypic interactions among cells into an organism are really crucial factors in cancer pathogenesis.

The fact that science has not yet developed a full understanding of the rules governing the phenomenon of immunotolerance is a huge obstacle to the comprehension of cancer immunology.

Today, it is commonly understood that much of the mechanisms of immunotolerance are achieved in the thymus gland and bone marrow during embryonic and early postnatal development, where the B and T lymphocytes, that have developed immunological reactivity to the body's normal proteins, are eliminated or functionally inactivated. Later, lymphocytes circulating throughout the body develop "peripheral tolerance" to proteins that they have encountered in tissues distant from the thymus and bone marrow. Once again, tolerance is probably achieved by eliminating lymphocytic clones that recognize self tissue antigens.

Recently, and most importantly, the role of Tregs have been indicated as a crucial element in establishing of immunotolerance. Tregs can directly inhibit and even kill both CTLs and THs that recognize the same antigens. They are usually responsible for taking at bay the immune system against self-tissues, in order to prevent autoimmune reactions.

Tregs alterations that are investigated for their role in autoimmune diseases, seem to also play a major role in immunoevasion of cancer cells. In normal individuals, the Tregs represent only 5 to 10% of the global population of CD4+ lymphocytes, the remainder being THs. In cancer patients, however, this number is often significantly increased. In addition, Tregs have been found, in large numbers, among the tumor-infiltrating lymphocytes (TILs) present in lung, ovarian, breast, and pancreatic carcinomas, as well as in tumor ascites.

The exact mechanisms by means of which Tregs are recruited and converted into tumor defenders are unknown. A simplistic possibility is that tumors release of chemokines that could constitute a molecular recall for these immunocytes on their surface. Once present within tumor masses, Tregs can suppress the actions of THs that are instrumental in mobilizing both the humoral and cellular arms of the adaptive immune response, that would be otherwise fully competent in attacking and killing tumor cells.

The existence of Tregs clearly destabilizes many conclusions drawn in the past years concerning the role of TILs in the pathogenesis of cancer. They have been formerly assumed to be cytotoxic cells that are actively involved in eliminating cancer cells. In fact, correlations of the prognosis of a number of carcinoma patients with the presence of substantial numbers of TILs in their cancers evidenced that patients lacking significant populations of TILs had a poorer prognosis. However, the study of the relative composition of TILs subfamilies might reveal yet further types of stromal cell recruitment into the tumor mass in order to support its expansion.

Also macrophages, traditionally depicted as front-line soldiers of the immune response can become allied of tumors. Macrophages usually deal with infectious agents, such as bacteria, by phagocytising them and then instructing the long-term immune response through antigen presentation by MHCII membrane proteins. A subset of macrophages has been observed to detect and kill cancer cells. Yet other subsets clearly fail to do so in many types of tumors. These are tumor associated macrophages (TAMs) that are crucial helpers of cancer cells in invasion and metastasis.

These cells are thought to be continuously recruited in tumors, attracted by the inflammatory status of these zones. They may even initially attempt to fight, but they are usually unsuccessful since carcinogens cannot be removed because they are particularly resistant infectious agents or inorganic compounds, such as asbestos particles, against which immunocytes have very few power. Alternatively, they could be weakened by the action of a series of immunosuppressive factors released by cancer cells or their alleys. These 'frustrated' immunocytes are thought to undergo a process of maturation or 'education' within similar tumor microenvironments, resulting in drastic phenotypic changes, characteristics of TAMS.

The tumor microenvironment causes macrophages to suppress their functions as immune cells and to adopt trophic roles, as it happens physiologically during development and repair of tissues in wound healing. The switch of innate immune cells toward being pro-tumorigenic can be achieved in many ways most of which have not been yet clearly defined. One way, this can occur by altering the intercellular crosstalk through peculiar cytokines levels in the tumor microenvironment, a strategy known to also significantly dampen the antigen presenting function of other immune cells; in particular dendritic cell (DCs), which are today considered the most important antigen-presenting cells for modulating anticancer immunity responses. The subject is in fact considered so important that it is most widely treated in the literature and even various DC based cancer vaccines are under development, in an attempt to help the immune system reacting against clinically relevant tumors.

In the early stages of tumorigenesis, macrophages are found at sites along the basement-membrane breaks or at the invasive front of more advanced tumors. This suggests that tumors could exploit the normal matrix remodelling capacities of macrophages, thus acquiring the ability to migrate through the surrounding stroma. Multiphoton imaging microscopy has provided images revealing remarkable interactions between tumor cells, macrophages, and blood vessels.

Macrophages play a role in the initiation of angiogenesis but have a role also in the remodelling of the vasculature once formed, to realise a coherent vascular flow. For similar functions, macrophages appear to be used in wound healing and during the remodelling of the vasculature in the eye during postnatal development.

At a molecular level, the expression of MMPs from macrophages at sites of invasion, much like that performed by CAFs seen before, causes the release digestion of the extracellular matrix and the release of VEGF molecules formerly bound there. This release, in turn, promotes neoangiogenesis at invasion sites. In humans, clinical evidences have been collected that show a correlation between local macrophage density and areas of intense angiogenesis defined by the presence of microvessels, confirming a role for macrophages in neoangiogenesis.

Only later, in tumor progression, collaboration among TAMs and cancer cells is elaborated and becomes a sort of symbiosis very destructive for the rest of the organism. Briefly, TAMs enhance tumor incidence, progression, and metastasis. This view has been confirmed by results in mice experiments, in which inhibition of metastasis was obtained when macrophages were purposefully ablated from the animals.

Cancer cells receive much aids functionality by macrophages, as well as from other alleys. What do these servants receive in exchange? Certainly various immediate rewarding, like closeness to privileged flows of nutrients, stimulations through mitogenic factors, and the company of other powerful stromal alleys. However, their decision of serving their wicked masters, in their endless search for personal pursuit, will reveal a short-sighted one, since it will doom an entire biological system to destruction.

Concluding Remarks

From the beginning of this book, we have described how cancer cells arise. Later and progressively, we have seen how cancers are composed of many different cell types, like normal tissues, albeit with qualitative and quantitative differences. We have discovered how intercellular relationships are completely abnormal when cancer advances and how altered heterotypic interactions among cells is a great contributor to the disease. Finally a more adequate level of detail in cancer biology and histology is emerging.

In front of such complex set of intercellular relationships, similar in its complexity to interrelationships among animals in ecosystems, we are beginning to appreciate how cancer cells find ways to raise. A complex network of information exchanges, of synergies and conflicts among cancer cells and stromal cells, immunocytes and many others, is in place during the escalation of the pathology. From these evidences a series of conclusions could stem, some of which, with the character of multidisciplinary are left for later and some merely technical but probably worth mentioning at the end of this chapter.

In particular, the awareness of heterotypic interactions in oncology is resulting in a debate on the inadequacy of current models used in experimental settings in cancer research. In fact, current oncology models often rely on implanting human tumor cell lines subcutaneously in immunocompromised mice. While being easy to realise and enabling to readily monitor the growth of tumors, these implants produce quite artificial situations, resulting in microenvironmental assets very unlikely to occur in real neoplasias.

Even more markedly, the ongoing preclinical experimental settings, involving isolated cancer cell cultures, completely missing all the universe of heterotypic interactions in real tumors, should have very poor capabilities of predicting the efficacy of experimental therapeutics against natural occurring tumors. The inadequacies of currently available preclinical models, and the resulting inability to accurately predict clinical outcomes in humans, is costing too much in term of money and, most importantly, in patient's lives. A plausible solution could be the use of real cancers in laboratory animals, spontaneous or induced in vivo by carcinogens. Similar models, despite being much slower than those currently adopted, could be a transitory solution, waiting for more powerful preclinical models to be validated.

Further Readings

Balkwill, F., Cancer and the chemokine network. Nat Rev Cancer, 2004. 4(7): 540–50.

Bergers, G. and L.E. Benjamin, Tumorigenesis and the angiogenic switch. Nat Rev Cancer, 2003. 3(6): 401–10.

Bhowmick, N.A., E.G. Neilson, and H.L. Moses, Stromal fibroblasts in cancer initiation and progression. Nature, 2004. 432(7015): 332–7.

Bissell, M.J. and D. Radisky, Putting tumours in context. Nat Rev Cancer, 2001. 1(1): 46–54.

Braun, S., et al., Cytokeratin-positive cells in the bone marrow and survival of patients with stage I, II, or III breast cancer. N Engl J Med, 2000. 342(8): 525–33.

Chabner, B.A. and T.G. Roberts, Jr., Timeline: Chemotherapy and the war on cancer. Nat Rev Cancer, 2005. 5(1): 65–72.

Clevers, H., At the crossroads of inflammation and cancer. Cell, 2004. 118(6): 671–4.

Condeelis, J. and J.W. Pollard, Macrophages: obligate partners for tumor cell migration, invasion, and metastasis. Cell, 2006. 124(2): 263–6.

Condeelis, J. and J.E. Segall, Intravital imaging of cell movement in tumours. Nat Rev Cancer, 2003. 3(12): 921–30.

Coussens, L.M. and Z. Werb, Inflammation and cancer. Nature, 2002. 420(6917): 860–7.

Deeb, K., D. Trump, and C. Johnson, Vitamin D signalling pathways in cancer: potential for anticancer therapeutics. Nat Rev Cancer, 2007. 7(9): 684–700.

Derynck, R., R.J. Akhurst, and A. Balmain, TGF-beta signaling in tumor suppression and cancer progression. Nat Genet, 2001. 29(2): 117–29.

Dranoff, G., Cytokines in cancer pathogenesis and cancer therapy. Nat Rev Cancer, 2004. 4(1): 11–22.

Ferrara, N., VEGF and the quest for tumour angiogenesis factors. Nat Rev Cancer, 2002. 2(10): 795–803.

Gilbertson, R. and J. Rich, Making a tumour's bed: glioblastoma stem cells and the vascular niche. Nat Rev Cancer, 2007. 7(10): 733–736.

Hanahan, D. and J. Folkman, Patterns and emerging mechanisms of the angiogenic switch during tumorigenesis. Cell, 1996. 86(3): 353–64.

Hanahan, D. and R.A. Weinberg, The hallmarks of cancer. Cell, 2000. 100(1): 57–70.

Huang, P., et al., Superoxide dismutase as a target for the selective killing of cancer cells. Nature, 2000. 407(6802): 390–395.

Jain, R.K., Normalization of tumor vasculature: an emerging concept in antiangiogenic therapy. Science, 2005. 307(5706): 58–62.

Joyce, J.A., Therapeutic targeting of the tumor microenvironment. Cancer Cell, 2005. 7(6): 513–20.

Kalluri, R., Basement membranes: structure, assembly and role in tumour angiogenesis. Nat Rev Cancer, 2003. 3(6): 422–33.

Karin, M., et al., NF-kappaB in cancer: from innocent bystander to major culprit. Nat Rev Cancer, 2002. 2(4): 301–10.

Kenny, P.A., C.M. Nelson, and M.J. Bissell, The ecology of tumors. The scientist, 2006. 20(4): 30.

Kerbel, R. and J. Folkman, Clinical translation of angiogenesis inhibitors. Nat Rev Cancer, 2002. 2(10): 727–39.

Lin, E.Y. and J.W. Pollard, Tumor-associated macrophages press the angiogenic switch in breast cancer. Cancer Res, 2007. 67(11): 5064–6.

Mueller, M.M. and N.E. Fusenig, Friends or foes – bipolar effects of the tumour stroma in cancer. Nat Rev Cancer, 2004. 4(11): 839–49.

Mundy, G.R., Metastasis to bone: causes, consequences and therapeutic opportunities. Nat Rev Cancer, 2002. 2(8): 584–93.

Murphy, P.M., Chemokines and the molecular basis of cancer metastasis. N Engl J Med, 2001. 345(11): 833–5.

Nickoloff, B., Y. Ben-Neriah, and E. Pikarsky, Inflammation and Cancer: Is the Link as Simple as We Think? J Investig Dermatol, 2005. 124(6): x–xiv.

Passegue, E., Cancer biology: a game of subversion. Nature, 2006. 442(7104): 754–5.

Pollard, J.W., Tumour-educated macrophages promote tumour progression and metastasis. Nat Rev Cancer, 2004. 4(1): 71–8.

Potter, J., Morphogens, morphostats, microarchitecture and malignancy. Nat Rev Cancer, 2007. 7(6): 464–474.

Ridley, A., Molecular switches in metastasis. Nature, 2000. 406(6795): 466–7.

Schwartz, M.A. and M.H. Ginsberg, Networks and crosstalk: integrin signalling spreads. Nat Cell Biol, 2002. 4(4): E65–8.

Soucek, L., et al., Mast cells are required for angiogenesis and macroscopic expansion of Myc-induced pancreatic islet tumors. Nat Med, 2007. 13(10): 1211–1218.

Walkley, C.R., et al., A microenvironment-induced myeloproliferative syndrome caused by retinoic acid receptor gamma deficiency. Cell, 2007. 129(6): 1097–110.

Weigelt, B., J.L. Peterse, and L.J. van 't Veer, Breast cancer metastasis: markers and models. Nat Rev Cancer, 2005. 5(8): 591–602.

Whitworth, P.W., et al., Macrophages and cancer. Cancer Metastasis Rev, 1990. 8(4): 319–51.

Wiseman, B.S. and Z. Werb, Stromal effects on mammary gland development and breast cancer. Science, 2002. 296(5570): 1046–9.

Beyer, M. and J.L. Schultze, Regulatory T cells in cancer. Blood, 2006. 108(3): 804–11.

de Visser, K.E., A. Eichten, and L.M. Coussens, Paradoxical roles of the immune system during cancer development. Nat Rev Cancer, 2006. 6(1): 24–37.

Dunn, G.P., et al., Cancer immunoediting: from immunosurveillance to tumor escape. Nat Immunol, 2002. 3(11): 991–8.

Gilboa, E., The promise of cancer vaccines. Nat Rev Cancer, 2004. 4(5): 401–11.

Houghton, A.N., H. Uchi, and J.D. Wolchok, The role of the immune system in early epithelial carcinogenesis: B-ware the double-edged sword. Cancer Cell, 2005. 7(5): 403–5.

Karin, M., T. Lawrence, and V. Nizet, Innate immunity gone awry: linking microbial infections to chronic inflammation and cancer. Cell, 2006. 124(4): 823–35.

Nagaraj, S., et al., Altered recognition of antigen is a mechanism of CD8+ T cell tolerance in cancer. Nat Med, 2007. 13(7): 828–35.

Sakaguchi, S., Naturally arising Foxp3-expressing CD25+CD4+ regulatory T cells in immunological tolerance to self and non-self. Nat Immunol, 2005. 6(4): 345–52.

Schwartz, R.H., Natural regulatory T cells and self-tolerance. Nat Immunol, 2005. 6(4): 327–30.

Turk, M.J., et al., Multiple pathways to tumor immunity and concomitant autoimmunity. Immunol Rev, 2002. 188: 122–35.

Zitvogel, L., A. Tesniere, and G. Kroemer, Cancer despite immunosurveillance: immunoselection and immunosubversion. Nat Rev Immunol, 2006. 6(10): 715–27.

Chapter 5
Selfish Business

Evil when we are in its power is not felt as evil but as a necessity, or even a duty.
Simone Weil, Gravity and Grace, 1947.*

Punishing the Greedy: Punishment in Hell for the Deadly Sin of Greed is to be Boiled Alive in Oil. The image appeared in 1496 in Le grant kalendrier des Bergiers, published by Nicolas le Rouge in Troyes, France

---

* From www.famousquotes.com

Introduction

In the previous chapter, we have had a number of occasions for stressing an important concept in life science: that evolved life forms are constituted by a synergy of countless biological functions performed by various cell types and their forming molecules, in amazing equilibrium among each other. We have seen already as this wonderful status of equilibrium is perturbed during carcinogenesis, when a bunch of cells begin to act outside this former harmony. In this chapter we will deal with greediness as referred to cancer cells and their alleys. Neoplastic cells polarise energy and food supplies to their own use; giving back only waste products, killing the biological system in which they reside.

Cancer cells greedy behaviour has been previously traced back to one common peculiarity of oncogenes activity, that of stimulating the cell cycle in the absence of instructional signals normally imparted by other cells, thanks to their ability to enter and to progress through the cell cycle in a cell-autonomous way. This prerogative is fundamental for maintaining such a status of independency over the system. Mitogenic activated pathways are interconnected with those that regulate cellular nutrient uptake and metabolic activity. Rough calculations of the energy demands in cells indicate that they would very soon die of starvation when carrying alterations in the growth signalling pathways, without supportive alterations in their energetic management. Cells become autonomous for growth and proliferation would readily consume themselves without any danger for the whole organism. In an experimental verification of this view, transfection with the oncogene *myc* results in apoptosis of target cells secondarily to an abrupt start in transcription, translation and excessive consumption of adenisine triphosphate (ATP).

Cell survival is compatible with mutations that enhance fuel signalling alone, even in the absence of a specific mitogenic over-stimulation. An increased availability of nutrients would fuel increased ATP production and, as the cell takes up more energy than it needs and metabolizes it, mitochondria would generate reactive oxygen species (ROS) and other dangerous mutagenic chemical species in excess, that are pro-tumorigenic. It could even be possible, albeit not experimentally verified thus far, that alterations in metabolic pathways could constitute a predisposing condition for cells to become cancerous.

In cancer tissues, a wasteful and greedy utilisation of nutrient and energetic supplies can be realised because of a peculiar ability of cancer tissues to attract nutrients from the circulation. Empirically, that fact is easily verifiable, since cancer tissues avidly incorporate the glucose analogue fluoro-deoxy-glucose (FDG), an analogue of glucose, to the extent that this tracer is commonly used in positron emission tomography (PET) scans, the most sensible technique to diagnose tumors in vivo. This peculiar ability, in focalising wealth in their surrounding, could constitute an irresistible charm for other cell types and one of the reasons, actually even the main one, for we assist to those 'criminal associations' seen at work in the previous chapter. Hopefully, this ability could even constitute an Achil's heel to which aim at with anticancer therapies.

Greediness

The best documented manifestation of excessive and wasteful exploitation of nutrients in cancer tissues is that of glucose. Therefore we will focus on greediness for glucose, as a paradigm for greediness of nutrients in general.

Avidity for glucose has been explained in molecular terms. It has been initially tentatively explained in term of the so called Warburg effect. In brief, the fact that many types of cancer cells perform glycolysis at elevated rates, even in the presence of oxygen.

Oxidative phosphorylation in the mitochondria and glycolysis in the cytosol are two major metabolic pathways, by which ATP may be generated from glucose in the cell. If an impairment of the respiratory chain force the cell to rely on glycolysis for energy production, because the production of ATP is much more efficient through oxidative phosphorylation (36 ATP per glucose molecule) than by glycolysis (only 2 ATP per glucose molecule), a substantial increase of glycolytic activity would require a lot more of glucose to produce the same amounts of energy.

As early as in the 30s, the Nobel laureate Otto Warburg observed defects in the enzymatic activity related to the tricarboxilic acids (TCAs) cycle in cancer cells used in his experiments. He assumed that cancer was actually caused by defects in the oxidative phosphorylation in the mitochondria, forcing the cell to revert to glycolysis. In Warburg's view, the damage to the oxidative phosphorylation was the cause for cells to become undifferentiated and cancerous.

In a modern understanding of his theory, mitochondrial enzymes damage is considered a consequence of mitochondrial DNA (mtDNA) damage inflicted by mutagenic substances. Studies have revealed that mtDNA is, in fact, subject to high rates of mutations in cancer cells. Because the mitochondrial genome encodes a dozen of important protein components of the respiratory chain, mutations in mtDNA are likely to compromise the function of the respiratory chain. This damage to mtDNA can therefore be the primary cause of the increased metabolism in cancer cells but the increased metabolism can secondarily induce an overproduction of further mutagens, such as ROS, inducing a 'vicious cycle' in energy management. Damaged mitochondria are a promising issue for anticancer research and drugs that can hit altered mitochondria in cancer cells are one of the last frontiess in anticancer therapeutics research.[1]

In recent reports, however, it has been observed that many types of tumor cells could have a substantial reserve capacity to produce ATP by oxidative phosphorylation. Therefore, these new data support the common view that the high rate of glycolysis exhibited by most tumors is an adaptive response required to support cell growth rather than a compensation for defects in mitochondrial function. In a

---

[1] The study of Michelakis' group (Bonnet et al. in further readings), for instance, has famously shown that when mitochondrial impairment can be pharmacologically unlocked, in that case with the drug sodium dicholoroacetate (DCA), cancer cells can self-eliminate by apoptosis, leading to interesting reduction in the tumoral pathology.

modern formulation of the adaptive hypothesis, the role of hypoxic microenvironments, of abnormal expression of metabolic enzymes, and recently the contribution
of mitogenic pathways and specific oncogenes have been proposed.

In solid tumors, when the mass reaches a critical size and oxygen penetration
in the inner space becomes limited, hypoxia can be the condition that force cancer
cells to switch to the glycolytic pathway for ATP production, especially in those
cells that reside far from blood capillaries. Therefore, the increase in glycolysis rates
in cancer cells may be an adaptation to hypoxic microenvironments. In fact, under
such conditions, oxidative phosphorylation could not proceed because of insufficient
oxygen, even if the mitochondria were intact.

The milestone studies of Greg Semenza and colleagues have shown that the cellular response to hypoxia is generally controlled by the hypoxia inducible factor 1
(*HIF-1*) gene and its crucial gene product is responsible for the secondary activation
of the expression of many target genes, involved in disparate biological functions
such as angiogenesis, glucose uptake, glycolysis, growth factor signalling, apoptosis, and even cell motility and invasiveness.

The regulation of adaptive response of cancer cells is hardly exclusive through
HIF-1. Other proteins, such as the many oncogene products can drive the glycolysis
in the absence of *HIF-1*. In fact, early studies in rodent cells have shown that transfection with *ras* or *src* oncogenes led to an immediate marked increase in the glucose uptake, accompanied by an increase in the expression of glucose transporters.
Transformation of cells by the telomerase *TERT*, or by oncogenes, such as *SV-40T*
and *h-ras,* cause an increase in glycolysis dependency. Another oncogene, *bcr-Abl*
has also been implicated to play a role in glycolysis rates increase and inhibition of
bcr-Abl product reverse the Warburg effect observed in leukaemia cells.

Recent studies of Kraig Thompson and colleagues have indicated the AKT kinase
signalling pathway as a crucial player in this game of nutrients and energy requirements in cancer cells for selfish growth. In both the metabolic and cancer research,
AKT activation has been shown to be necessary and sufficient to regulate cellular
glucose uptake and constitutively active AKT is sufficient to stimulate glycolysis.
If an activated AKT is introduced into non-transformed cell lines in culture, cells
spontaneously switch over to aerobic glycolysis (the Warburg effect).

AKT contributes beyond glucose uptake to support tumor cell adaptation. AKT
phosphorylation of the enzyme ATP-citrate lyase (ACL), for instance, can result in
glucose-dependent lipid synthesis. Another AKT contribution to tumor cell adaptation occurs when high glucose input and AKT activation, together, inform the
increased uptake of amino acids necessary for cell growth. In this process, AKT
acts as the negative regulator of two tumor suppressors, TSC-1 (tuberous sclerosis
complex-1) and TSC-2, by degrading them, which negatively regulates target of
rapamycin (TOR), the master regulator of protein translation in mammalian cells.
AKT spontaneously activates TOR by at least other three independent mechanisms:
increasing ATP/ADP ratio by boosting glucose metabolism, increasing saturated
lipids to activate the lipid binding domain of TOR, turning off the negative repres-
sors of TOR, the most important mechanism of the three to fuel the cell with amino
acids.

To summarize, AKT pathway results in glucose uptake, fuelling glucose through mitochondria to produce lipids, amino acid uptake catabolism through TOR-dependent regulation of translation, for an increased protein synthesis. In all studies to date, this pathway is the controller of cell size and growth. However, although the synthesis of proteins and lipids account for how cells get bigger, there still remains a gap in our knowledge on how the switch in metabolism activate the replication of the genome. This would be a real turning point in the study of carcinogenesis that today, to the best of our knowledge, is still to be clearly proved.

Inside the adaptive hypothesis could fall also a model in which the aerobic glycolysis is accounted for in terms of modified versions of normal metabolic pathways. In this view, a modified version of the TCA cycle is proposed that would provide the best compromise between energy and building blocks requirements in growing cells, in particular the need of lipids to synthesise biological membranes. Scientists who have proposed this theory observe that the synthesis of one mole of a typical fatty acid requires the utilisation of ten moles of glucose. Therefore, it takes a huge amount of glucose to fulfil the lipogenic requirements of proliferating tumour cells. Neither anoxic glycolysis that produces only lactic acid nor normoxic glycolysis that completely oxidises glucose, regardless of the bioenergetics consequences of each, is compatible with the lipogenic requirements of tumour cells.

The ability of tumors to produce sufficient quantities of cytosolic acetyl-coenzyme A (acCoA) to fuel the synthesis of lipids needed for novel membrane and lipoproteins would depend on continued pyruvate degradation by the mitochondria, resulting in mitochondrial acCoA availability that is combined with oxaloacetate form citrate. However the citrate would be transported in the cytosol, where the enzyme ATP-citrate lyase (ACL) can convert citrate into acCoA and oxaloacetate. The oxaloacetate would be returned back to the mitochondria to complete a truncated citric acid cycle, whereas the remaining acCoA would provide a glucose-derived substrate for fatty acid synthesis. This alternative pathway would provide an additional increased production of ATP (12 ATP/glucose) compared to anoxic glycolysis (2ATP/glucose) to meet the bioenergetics requirements of the malignant proliferative activities of tumour cells. Therefore, it appears the best compromise between energy demands and precursor availability to fit autonomous cell proliferation.

However, the accelerated rate of conversion of glucose to pyruvate can exceed the rate of pyruvate oxidation. Accumulating pyruvate is subjected to reduction to lactate and the accumulation of this waste product, as well as many other sub products of the metabolism, contribute, together with other catabolic anomalies, to determine poisoning and acidification of the tumor microenvironment. The acidic tumor microenvironment, associated with accumulation of lactate secondary to increased glycolysis, is particularly selective for cancer cells with high survival capacity and malignant behaviours. The process is unsustainable in terms of a correct economy and biochemical balance in an organism. It is the host organism, not the parasitic tumour cell that suffers the consequences of this excessive (wasteful) utilisation of glucose. And for what matters the cancer cell, this is none of its business, at least in its short term perspective.

Because lipogenesis is essential for the growth/proliferation of tumour cells, those cells that do not 'glycolyse' must possess an alternative pathway to fulfil their proliferating demands. In fact, many malignant cell lines, as well as some normal cells, do not have an absolute requirement for glucose per se. Some author has proposed and provided evidence for the possible involvement of a glutamate involving pathway for lipogenesis. In fact, the two major products of the glutamate oxidation are citrate and alanine that are useful building blocks for further biosynthesis.

Extending the above observation, we must underline how this observation underlines how it is a mistake to expect that all tumour cells must exhibit a universal adaptive intermediary metabolism. The use of an elected variety of tumour cells to establish the general behaviour of tumour cells has lead often to imprecise arguments. In general, it is just the avidity for nutrient substrates that constitute a peculiar trait univoquely linked to malignancy and hopefully a target for therapeutic interventions.

Long Reach

In cancer cells, as a consequence for nutrient demands, membrane proteins that capture nutrients from the environment are usually overexpressed and work incessantly, at their maximum speed, in order to keep fuelled the intracellular space. This argument is usually used to explain the incredible affinity of cancer for nutrients. Recently, scientists have come to believe that there should be more, in addition to the above mentioned mechanism. In particular, a strong contribution to that ability could be provided by interaction between a peculiar electrical status of cancer cells and nutrient molecules.

The fact that cancer cells, as well as stem cells and proliferating cells, exhibit an electrical status very different from that of resting cells is a fact confirmed by number of experimental evidences in cell biology and pharmacology. This peculiar electrical status is realized through an altered expression or an increased activity of ion channel pores and transmembrane transporters.

Number of studies have been performed during the last decade on cancer cells in vitro under various physiological states. Their membranes electric status, as determined by ion channel conductance measured with the patch clamp technique, has been correlated with many abnormal functionalities linked with the pathogenesis of cancer. Various ion channel activities have been shown to be altered during transformation, proliferation, calcium (Ca^{2+}) intracellular signaling, regulation of pH and cell volume, motility, invasiveness, traffic of substrates and nutrients, multidrug resistance (MDR) and cancer cell interference with cellular and molecular components of immunity.

A predominant role in the literature is played by works addressing the role of potassium (K^+) channels in cell proliferation. Apoptosis, initiated by death-promoting molecules, is correlated to an early activation of K^+ currents and decrease in intracellular K^+ concentrations. For instance, growth- and mitosis-related enzymes require a minimal K^+ concentration and a loss of K^+ will reduce the proliferative activity. In fact, all these enzymes are controlled by the intracellular K^+ concentration. In general, it is probably safe to affirm that activation of K^+ conductances must stay within a certain conductance range to support proliferation; otherwise programmed cell death is triggered. The environmental conditions in which channel activation takes place, along with the magnitude of the activated conductance, essentially determine whether it supports proliferation or apoptosis.

An endless list of ionic alteration, documented in various texts, would be too long or elaborated for the scope of this text. However, it must be emphasized that these various altered ionic statuses constitute a source of peculiar interaction with electrically active species present in the organism. Glucose, aminoacids, fatty acids as well as many molecules of biological interest could be reached through long range interactions mediated by dielectric layers of water. In other words, the peculiar status of cancer cells could reveal be extremely important for their magnetism for nutrients, dealt with in this chapter.

In addition to these electrophysiology studies, electric and electromagnetic properties (EM) of cancer tissues have been tentatively measured. In very preliminary and rudimental experimental settings, not only constant electric fields but also electromagnetic waves generated by laser sources have been used to tentatively interact with cells. Preliminary results seem to indicate that very low frequency pulsed electric fields (in the 10 Hz range) have a major influence the ion membrane transport. Higher frequencies are not in resonance with the frequency of work of ion channels and would thus not be specifically active on cell electrophisiology.

EM based approaches are only at the very beginning. The adsorption of radiowaves has been used to differentiate between hyperproliferative and inflammatory tissues. Instruments that can perform this task are already into the market and in course of validation for clinical use. On a therapeutic scale, only a few applications have been demonstrated useful and feasible thus far. Wound healing process, similar for certain aspects to cancer as seen previously, has been favourably influenced by the application of EM impulses. Provided that scientists could be able to develop effective approaches in this field, it would be much fascinating to be able in the future to modulate the activity of biological systems without the need of introducing foreign drugs into the body. For the sake of truth, at present, this is more or less in the realm of science fiction.

Targeting Cancer Cell Avidity

In this book, I have voluntarily dismissed countless opportunities to enter into the dispersive and tricky terrain anticancer strategies each time we presented a particular biochemical mechanism. At this point, however, I have to give in because I think that an hopefully interesting new rational has been delineated: the targeting of cell greediness for nutrient supplies. The appeal of this approach appears to me particularly valuable, at least in theory, because it aims not only at classically intended cancer cells but also at all their greedy commensals, allied cells that constitute the tumor mass in its complexity.

For the sake of truth, albeit under a different theoretical frameworks, the rational is not new and attempts in this field are already ongoing. In case of PET scan positive cancers, for instance, molecules such as the glucose analogue 2-deoxyglucose (2-DG) or the FDG itself,[2] being molecules that do not produce ATP when metabolised, have been shown to induce starvation of cancer cells. This starvation could readily lead to death since these cells are usually highly addictive to and dependent on excessive glucose amounts. These preliminary attempts, despite having provided interesting preliminary results, have been shown rather weak in randomised phase III human clinical trials, despite being helpful in combination with classical chemotherapeutic regimens.

In this area of study, inhibitors of the glucose managing intracellular enzymes, mainly those involved in the glycolysis have been shown to cause depletion of ATP in cancer cells. Their effect is especially severe in cells with mitochondrial DNA defects. Inhibition of glycolysis with compounds such as for instance 3-bromopyruvate (3-BrPA), effectively kills colon cancer cells and lymphoma cells in hypoxic environment, in which cells exhibit high glycolytic rates and a decreased sensitivity to anticancer agents usually employed in the clinics including taxanes, anthracyclins, arsenic trioxide, antimetabolites, immunosuppressants and many others.

In vivo, the administration of this compound has led to notable results, such as complete remissions of hepatocellular carcinoma implants in mice. In these cases the effect could have been particularly potentiated for hepatic cells are very avid for pyruvate and its analogues. More in general, targeting the peculiar avidity of malignant cancer cells with poisoned baits, like toxic nutrient analogues, could finally hold the promise for an effective anticancer chemotherapy.

Recently, a series of optimised 3-BrPA derivatives, named glycolycins, are under development and hold great promises, even for other types of neoplasias. Inhibitors of glycolysis show effective anticancer activity in animal tumor models, suggesting that inhibition of glycolysis is a promising therapeutic strategy which may have broad clinical implications, even if the limitations in the predictive capacity of animal models currently employed also in these studies, which we discussed in the previous chapter, could hide tricky pitfalls.

In addition, there are potential concerns with the use of nutrient analogs and potential challenges that must be considered, when entering human clinical trials. In fact, certain normal tissues could use the same substrates and get intoxicated. Inhibition of glycolysis may be potentially toxic particularly to the brain, retinae, and testis. It is unclear whether these normal tissues can effectively use alternative energy sources (fatty acids, amino acids, etc.) to generate sufficient ATP through mitochondrial metabolism to support their cellular function, when the glycolytic pathway is inhibited during therapy. The field is open to intellectual contributions that could unlock actual problems and lead to better outcomes of anticancer therapies.

---

[2] FDG molecules with a non-emitting fluoride would be sufficient, because the emitting isotope is clearly unnecessarily costly, for this strategy.

Concluding Remarks

Because this book has been dedicated to cell selfishness, not because we had the intention of entering the discussion in cancer therapeutics, we have underlined one of the many hundreds of anticancer strategies under scrutiny in these days. Selfishness is a strong prerogative of cancer cells during their escalation to the power and death (actually together with the rest of a biological system they inhabit). Selfishness could hopefully lead them and their greedy commensals to exchange poisoned baits for another unnecessary rich meal.

Further Readings

Alirol, E. and J.C. Martinou, Mitochondria and cancer: is there a morphological connection? Oncogene, 2006. 25(34): 4706–16.

Arsham, A.M., et al., Akt and hypoxia-inducible factor-1 independently enhance tumor growth and angiogenesis. Cancer Res, 2004. 64(10): 3500–7.

Baysal, B., Mitochondria: more than mitochondrial DNA in cancer. PLoS Med, 2006. 3(3): e156; author reply e166.

Bonnet, S., et al., A mitochondria-K+ channel axis is suppressed in cancer and its normalization promotes apoptosis and inhibits cancer growth. Cancer Cell, 2007. 11(1): 37–51.

Cherubini, A., et al., HERG potassium channels are more frequently expressed in human endometrial cancer as compared to non-cancerous endometrium. Br J Cancer, 2000. 83(12): 1722–9.

Conti, M., Targeting K+ channels for cancer therapy. J Exp Ther Oncol, 2004. 4(2): 161–6.

Conti, M., Cancer determining information transmission and circulation. Cancer Metastasis Rev, 2007. 26(1): 215–20.

Conti, M., Targeting Ion Channels for New Strategies in Cancer Diagnosis and Therapy. Current Clinical Pharmacology, 2007. 2: 135–144.

Costello, L.C. and R.B. Franklin, 'Why do tumour cells glycolyse?': from glycolysis through citrate to lipogenesis. Mol Cell Biochem, 2005. 280(1–2): 1–8.

Dang, C.V., et al., The interplay between MYC and HIF in cancer. Nat Rev Cancer, 2008. 8(1): 51–6.

Degasperi, G.R., et al., Role of mitochondria in the immune response to cancer: a central role for Ca2+. J Bioenerg Biomembr, 2006. 38(1): 1–10.

Elstrom, R.L., et al., Akt stimulates aerobic glycolysis in cancer cells. Cancer Res, 2004. 64(11): 3892–9.

Fiske, J.L., et al., Voltage-sensitive ion channels and cancer. Cancer Metastasis Rev, 2006. 25(3): 493–500.

Galluzzi, L., et al., Mitochondria as therapeutic targets for cancer chemotherapy. Oncogene, 2006. 25(34): 4812–30.

Garber, K., Energy boost: the Warburg effect returns in a new theory of cancer. J Natl Cancer Inst, 2004. 96(24): 1805–6.

Garber, K., Energy deregulation: licensing tumors to grow. Science, 2006. 312(5777): 1158–9.

Gordan, J.D., C.B. Thompson, and M.C. Simon, HIF and c-Myc: sibling rivals for control of cancer cell metabolism and proliferation. Cancer Cell, 2007. 12(2): 108–13.

Harris, A.L., Hypoxia–a key regulatory factor in tumour growth. Nat Rev Cancer, 2002. 2(1): 38–47.

Ko, Y.H., et al., Advanced cancers: eradication in all cases using 3-bromopyruvate therapy to deplete ATP. Biochem Biophys Res Commun, 2004. 324(1): 269–75.

Kroemer, G., Mitochondria in cancer. Oncogene, 2006. 25(34): 4630–2.

Monteith, G., et al., Calcium and cancer: targeting Ca2+ transport. Nat Rev Cancer, 2007. 7(7): 519–530.

Pedersen, P.L., The cancer cell's "power plants" as promising therapeutic targets: an overview. J Bioenerg Biomembr, 2007. 39(1): 1–12.

Pelicano, H., et al., Glycolysis inhibition for anticancer treatment. Oncogene, 2006. 25(34): 4633–46.

Rustin, P., Mitochondria, from cell death to proliferation. Nat Genet, 2002. 30(4): 352–3.

Sarkisian, C.J., et al., Dose-dependent oncogene-induced senescence in vivo and its evasion during mammary tumorigenesis. Nat Cell Biol, 2007. 9(5): 493–505.

Semenza, G.L., Targeting HIF-1 for cancer therapy. Nat Rev Cancer, 2003. 3(10): 721–32.

Wang, H., et al., HERG K+ channel, a regulator of tumor cell apoptosis and proliferation. Cancer Res, 2002. 62(17): 4843–8.

Wang, Z.H., et al., Blockage of intermediate-conductance-Ca(2+)-activated K(+) channels inhibits progression of human endometrial cancer. Oncogene, 2007. 26(35): 5107–14.

Xu, R.H., et al., Inhibition of glycolysis in cancer cells: a novel strategy to overcome drug resistance associated with mitochondrial respiratory defect and hypoxia. Cancer Res, 2005. 65(2): 613–21.

Zamzami, N. and G. Kroemer, The mitochondrion in apoptosis: how Pandora's box opens. Nat Rev Mol Cell Biol, 2001. 2(1): 67–71.

Zhang, H., et al., HIF-1 inhibits mitochondrial biogenesis and cellular respiration in VHL-deficient renal cell carcinoma by repression of C-MYC activity. Cancer Cell, 2007. 11(5): 407–20.

Epilogue

"For I was hungry and you gave me something to eat, I was thirsty and you gave me something to drink, I was a stranger and you invited me in, I needed clothes and you clothed me, I was sick and you looked after me, I was in prison and you came to visit me... whatever you did for one of the least of these brothers of mine, you did for me."

The gospel of Matthew 25: 35,36,40.

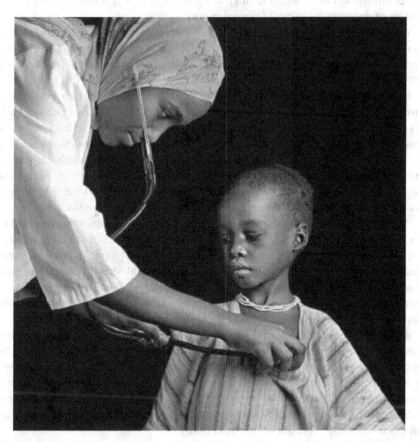

Nurse and child; jinka hospital; jinka, ethiopia (Picture in the cover of Novartis 2005 Corporate Annual Report.)

Selfish cells have been the subject of this work: why they raise, how they behaves, conquering and devastating the human body. Actually we went to observe that cancer is not solely dependent on those deranged cells, but characterised by a wider loss of ensemble harmony that normally keeps together all the various cells in living tissues. As a consequence, we have described how 'criminal associations' around neoplastic cells form niches of selfishness in the human body that grow at every cost, perpetrating waves of actual biological barbarian and devastation.

As it usually happens in our human lives, when facing violence and injustice, we are left with puzzling interrogatives, such as why these tendencies manage to prevail at times during history. Maybe we should be even more puzzled when considering that similar tragedies are consumed also in the microcosm by presumably naive biological entities that are cells, as the reality of cancer tragically attests.

Frankly, explanations for the escalation of such phenomena that could be provided in cold scientific terms appear quite short fetched. Even more unsatisfactory and contradictory appears the evolutionistic based explanation for selfishness, since cancer affirmation as a case of biological evolution is clearly a warning and a declaration of defeat of the cinical Darwinian perspective.

While researchers are keeping on advancing proposals, changing fighting strategies in the hope to reach better explanatory models and especially practical solutions for winning one day this terrible disease; as humans we could be tempted to think that we won't ever reach an ultimate answer to similar questions in this life.

However, I like to think that we should not be inclined to such a pessimism. Answers are available in many fields of human experience, the study of cancer biology being just a particularly explicit one. From human experience, at least one main answer can be extrapolated very clearly and generally: it is the absurd and unwise tendency to follow personal profit, losing the sense of relationship with our similar and with nature. At the opposite side we could meet the choice of love, of care for the others. Where would we be and where would we go without love? My wish is that men want to maintain reference to it, even in their future evolution.

At the end, let me apologize with the reader if most of the concepts presented in this book would be fast gone, overcome by the rapid expansion of scientific frontier in cancer research, really racing up in these days. Actually, one of my goals was exactly to help in that process. In any case, I like to think that this work has been an occasion for reflecting over important issues around our humanity, even a chance for a warning of respect and equilibrium among humans and between humans and nature. As humans, we strongly must believe nothing is in vain in our lives.

As a very final note, I would like to recommend a simple text, found in my wanderings on the internet. The author is not an expert in the field, at least she told me. Her advice I went to share even if not punctually or for the tones, but in main terms. I must confess I had something similar going in my mind but I never managed to put it down in words, as clearly as she does. I thank her for she has agreed to let me submit it to the judgement of the reader. (Appendix from http://www.mit.edu/~rei/spir-spiritcancer.html).

Appendix

Cancer Cells, Healthy Cells: Spirit, Society, and Reality

Cancer: The Analogy

The other day, worrying about a friend facing chemotherapy, I was riding the commuter rail and reading Peace Pilgrim: Her Life and Work in Her Own Words to try to gain a little spiritual peace. Between the worry about the friend who was fighting off cancer, and my ponderings about humanity, something clicked when I read this passage:

> "We begin feeling very separate and judging everything as it relates to us, as though we were the center of the universe. Even after we know better intellectually, we still judge things that way. In reality, of course, we are all cells in the body of humanity. We are not separate from our fellow humans. The whole thing is a totality. It's only from that higher viewpoint that you can know what it is to love your neighbor as yourself. From that higher viewpoint there becomes just one realistic way to work, and that is for the good of the whole. As long as you work for your selfish little self, you're just one cell against all those other cells, and you're way out of harmony. . .." (from Chapter , or "My Steps Toward Inner Peace")

> That's when I asked myself, are we cancer cells?

> Yes, I know, it's an old analogy. People refer to their social enemies or social problems as "cancers." But what does this really mean?

Cancer is, quite simply, our own cells that have turned so "selfish" that they no longer do their assigned roles (e.g. be a liver cell, be a colon cell, be a lung cell), and instead just start selfishly hogging nutrients, fooling nearby cells into helping them, multiply, and eventually destroy the body. Meaning, kill the person they were supposed to be helping.

Some years ago, I had a fascinating conversation about how social problems have similarities with cancer. According to the person I was talking with, the organs in our body normally have a constructively competitive relationship with each other. The brain competes for blood against the other organs, for example. The needs and demands of various organs play against each other, and sometimes in emergencies the brain and heart will demand all available oxygen and to heck with everyone else, but all in all, the body manages to regulate things for the good of the whole.

Cancer cells, in contrast, are destructively competitive: they compete, but they play a zero-sum game where if they win, everyone else (the rest of the body) loses. (Not that the cancer really "wins" – it dies when the body it was depending on dies.)

If we look around at society, we can see spiritual cancer sprouting up in just about every sector. On the streets, we can find gangs recruiting members, expanding their territories, disposing of enemies with violence that cares not a whit who might die, and spreading the drugs by which countless people will fall into addiction and despair (and wreak havoc on their own families). In the rarefied strata of big business, the widely-known corporate scandals point to powerful individuals who choose to work for personal gain over all else, sacrificing everything from ethics, to employees (if you've worked for a selfish and arrogant boss you know what I mean), to even the environment – and sometimes, the selfishness can even destroy the company, leaving innocent employees and investors holding the ruined pieces of their job and investment hopes.

Government, medicine, even science and religion – we will find examples of those who want to succeed or be proved "right" no matter the cost to anyone else.

The Similarities: Cancer vs. Social Cancer

Don't these "fallen" human beings display the appropriate signs of being cancerous cells in the body of society? They. . .

...had the potential to be productive, functioning members of society, just as the first cancer cell used to be a normal cell.

...may turn "bad" because of abuse or crime, just as cancer cells may be created when exposed to harmful chemicals or harmful radiation.

...forget the true role they were supposed to be fulfilling. Just as a cancerous lung cell forgets it was supposed to be helping the body, so the "fallen" politician (as an example) forgets he was supposed to be helping the people of his country – genuinely helping, long-term and not just short-term – instead of consolidating his power and feeding just his political buddies.

...seek selfish goals (money, power, recognition), just as cancer cells seek to merely eat well and multiply.

...corrupt and multiply by dragging their children and others nearby into the celebration of self-centeredness and arrogance and oppression, just as cancer cells multiply and form tumors.

...negatively inspire other people far away to do similar actions (e.g. copycat criminals) or escape justice and repeat the actions far away (e.g. the serial corporate CEO who ruins company after company), just as cancer cells can escape from the initial cancer and metastasize elsewhere.

...enjoy power and energy, just as cancer cells love and feed on glucose, the basic chemical energy unit of the body.

...fool others into thinking they are helpful and good, thereby winning votes, money, power, fame, fortune – just as cancer cells fool the body's defenses, and manage to call in new blood vessels and a steady supply of glucose and oxygen.

...waste tremendous resources, just as cancer cells "steal" their energy and are terribly inefficient about using that energy (according to one website, cancer cells have a metabolism eight times greater than normal cells). Corrupt officials, drug dealers, companies, etc. are those who destroy natural resources, induce people to consume wasteful products (think of how public transportation in LA was destroyed to help the auto industry), and suppress innovative ideas just so as to cling to established power and income. The resources that could feed and clothe the entire world get sucked into a few people's unquenchable and bottomless hunger for power, and so innocent millions starve and the world slowly dies. How is this different from a cancer that steals nutrients and starves the person to death?

...sometimes strive for absolute safety as well as immortality in some form, just as some cancer cells avoid normal cell death (apoptosis) by chemical and biological means.

...cause untold misery, just as cancer brings immense suffering.

...and, like a cancer cell that dies when its host dies, the human being who has turned cancerous will also die when the society he is destroying dies.

See some parallels? Our universe is one in which patterns repeat themselves. Mathematical patterns pop up in unexpected places, from the shapes of bee hives and Romanesco broccoli, to the golden ratio, to the shapes of flowers. We see how simple components (each with its own roles) aggregate to form complex components, which each have their own roles and yet aggregate to form even higher components: atoms form molecules, molecules build up cells, cells build up organs, organs make up bodies, and bodies form societies, and societies form worlds. In every level where the spark of animate life exists, where spirit most obviously holds sway, there are repeating patterns of life and behavior. So it is no surprise, then, that just as cells can go bad, human beings can go bad, and eventually entire companies, governments, and countries can start going bad.

The moment we decide that our needs are more important than anyone else's, and that our good must be ensured even at others' expense, then we have become a cancer. (Yes, it is true that if a choice must be made, the body will sacrifice all else to save the brain and the heart. But this does not excuse a brain that seeks its own pleasure at the expense of the rest of the body – such as those who choose mind-altering drugs that destroy health. And woe to the heart beating on after the rest of the body has shut down – it will soon die, too. Even losing a relatively "unimportant" organ like the gallbladder is now known to cause long-term problems for the rest of the body (such as cancer!). A body that cares only for its most glamorous parts, and not for the rest of itself, is heading for trouble.)

What Can Be Done?

So, if our society is suffering from cancer – as indeed it appears to be – what can be done about this?

The current way of fighting cancer is to cut it out (along with surrounding healthy tissues) and destroy it, then flood the body with toxic chemicals. I suppose the social equivalent is police action, war, prisons, executions, and confinement. How expensive this way is! At times it becomes absolutely necessary, but still, knowing the trauma and suffering of both war and, on a smaller scale, harsh medical treatments, perhaps it is fair to ask, "Is there a better way? A less violent way? A higher way?"

We have only lately started thinking about prevention. How much better to stop the cells from going bad, whether through better nutrition, proper exercise, or through reducing exposure to toxic chemicals. And likewise how much better it is to be building up a positive, caring society, in which all are cared for, and abuse is avoided and neutralized before tender young personalities are deeply scarred?

And finally, there is the possibility of change. When I first wrote this section, I thought that cancer cells could not possibly change back to normal cells, contrasting strongly with human beings who could choose to become a contributing member of society again.

But I was wrong. According to Dr. Jean-Pierre Issa of the M.D. Anderson Cancer Center, cellular changes may sometimes be reversed in epigenetic cancers (those in which the DNA is intact, but the expression of some genes is silenced through other mechanisms). Rather than killing cancerous cells, the treatment involves reactivating the genes that had been silenced. Then, if those genes succcessfully become operational again, the cell apparently changes back to normal – the results include a remarkable rate of remissions, or even a disappearance of the cancer.

And of course, we know that human beings can change as well.

Such cases are rare, but they exist. The murderer in prison who repents and changes his life; the greedy corporate leader who awakens to a new way of thinking; the angry gang leader who discovers that hatred and violence solve nothing, and only compassion and caring offer the real way out.

What can change these people from cancerous to caring, from malignant to magnificent?

Isn't it love?

What can stop people from becoming cancerous in the first place?

Isn't it love?

To continue the Peace Pilgrim quote from the top of this page:

"...But as soon as you begin working for the good of the whole, you find yourself in harmony with all of your fellow human beings. You see, it's the easy, harmonious way to live." (from Chapter , or "My Steps Toward Inner Peace")

This is not new wisdom. The theme is very similar to some astute text included in 1 Corinthians 12:

For the body is not one member, but many. If the foot shall say, Because I am not the hand, I am not of the body; is it therefore not of the body? And if the ear shall say, Because I am not the eye, I am not of the body; is it therefore not of the body? If the whole body were an eye, where were the hearing? If the whole were hearing, where were the smelling?[...]

And the eye cannot say unto the hand, I have no need of thee: nor again the head to the feet, I have no need of you. Nay, much more those members of the body, which seem to be more feeble, are necessary: And those members of the body, which we think to be less honourable, upon these we bestow more abundant honour; and our uncomely parts have more abundant comeliness. For our comely parts have no need: but God hath tempered the body together, having given more abundant honour to that part which lacked.

That there should be no schism in the body; but that the members should have the same care one for another. And whether one member suffer, all the members suffer with it; or one member be honoured, all the members rejoice with it.

Indeed, if one member suffers – if one person is enduring great pain – then we know now that eventually, he is likely to "fall away" and, in effect, turn cancerous. We know this from examples all around us of violence or greed. The entire body suffers when one member suffers. But if one unhappy, destructive person is returned to society, happy and healed, loved and loving both, then the entire body is the better for it. After all, every human being has a unique and individual aptitude, potential, and way to contribute!

And Whose Body Is It?

Romans 12:5:
So we, being many, are one body in Christ, and every one members one of another.

We may care about only ourselves, and do whatever it takes to please just us; we would have no allegiance to others, no compassion. We would be renegade, solo cells – no more than a single-celled organism. But most of us choose to belong to some body or other (recognizing also that groups can usually get more done than an individual). The bodies we choose tend to be small and familiar. We may identify most closely with our own family, and live and work for that family. We may ally most closely with our company, and work like mad for that company, and do everything we can for that company. Or, we may choose our country; we may fly our flag, cheer our leaders, work for, fight for, die for our country.

But belonging to these "bodies" is no guarantee of social good. In some cases, our family is a street gang, and our work is to sell drugs that causes misery that might cause others to sink into evil. In some cases, our company is greedy, and our work is to make profit no matter what, and the fruit of this practice might be poverty and environmental destruction. In some cases, our country is fascist; it may be persecuting minorities and waging war on its neighbors, and our work is to overcome and kill the enemy; and we already know the horrors that result from that.

How is this different from choosing to be part of a body that causes its cells – us – to become cancerous through the poor choices of the leaders? And how can we want to help spread cancer throughout the body we care about? Or, if the body we chose is getting fat at the expense of the rest of the world, do we really want to be part of what is essentially a tumor on the face of the planet?

So – as C. S. Lewis pointed out, we must choose for our body, and our leader, something higher. Preferably, we will choose the highest concept of good that we know. Perhaps we will even choose to follow God.

But in some cases, our God is a god of narrowness and condemnation, and our work may be to destroy all those who disagree with us, even, perhaps, through terrorism. Don't we need to choose our God wisely, measuring our God against the highest principles we know? And thus don't we already know that our God must be a God of love?

The Body of Love

Suppose we believe Christ Jesus is a good leader, a teacher of love, and that he was sent by God to bring humanity to God. Suppose we want to join that oft-quoted "body of Christ" thing.

If the Christ that we choose likewise loves all of us as much as (if not more than) he loves himself, doesn't that mean he will take care of us far better than we take care of our mortal bodies? This is good; that means that he will take care of us, heal us, feed us, and certainly keep us from becoming cancerous.

If the Christ that we choose even loves those that we have condemned, or judged, or hate, what does that mean? Could that mean that everyone is potentially part of the same body? Theologists may argue that one must be born again to be part of Christ's body, and all who turn away are forfeit, but didn't Jesus teach us to love everyone, including our enemies? Didn't he teach us not to judge and not to condemn?

Why would God ask us to love that which is unloveable? If we are expected to love everyone, then everyone must be loveable (God doesn't ask us to do impossible things). We are not to judge who is or is not part of the same body. And supposing that "sinner" over there really is separate from me. Doesn't love bind together, and make whole, and thereby unite me with that "sinner" over there and erase the separation? Doesn't love force us to view even those we thought were "impossible" or "unsaveable" or "unloveable" as worthy of care, just as every part of a body is worthy of care for the sake of the whole? Doesn't love show how we really are connected, no matter how we may try to separate ourselves or believe we are "different," just as the parts of a body are connected and each part is specialized, and harm to one harms all? And doesn't the act of loving others force us to abandon our selfishness and arrogance, those very things that characterize cancer?

The Ultimate Solution

Can we see that viewing others through love, through the eyes of God, unifies all of us into one body? And that the action of loving dissolves the boundaries we thought were there, sends caring to the malnourished, sends healing to the bitter and twisted, uses forgiveness to return renegades back to their loved ones, and transforms the illusion of separation and "us vs. them" into the truth of wholeness and unity?

Is this, perhaps, how bodies are miraculously healed of cancer?

And isn't this how human beings are miraculously healed of spiritual cancer?

The best preventative, the best medicine, the best cure for cancer of the world is love, and we (you, me) are precisely the ones who need to practice the love. We can't just want to get that love; it is in the giving that we turn ourselves away from becoming a cancer, and help others do the same, and bring wholeness and healing and unity to all.

The key is to start loving. The key is to give, rather than get. The key is to practice love right now.

Then we become a network of healthy cells, cells that not only are healthy, but which spread health throughout the body. And the body becomes stronger and healthier. . .

. . .and just as a person freed from cancer discovers health and energy and is given a new lease on life . . . imagine what a cancer-free, loving, giving world would be like!

I can't wait.

Only love lets us see progressively higher: the cells unified into organs, the organs unified into people, the people unified into societies, the societies unified into nations, the nations unified into worlds, the worlds unified into creation, and all creation unified in God. No more separation – no more strife – no more war – all becomes one in God. If we choose love.

Index